AF731751

SURPLUS - 2
LIBRARY OF CONGRESS
DUPLICATE

SURPLUS - 1
LIBRARY OF CONGRESS
DUPLICATE

CATALYTIC INCINERATION FOR CONTROL OF VOLATILE ORGANIC COMPOUND EMISSIONS

CATALYTIC INCINERATION FOR CONTROL OF VOLATILE ORGANIC COMPOUND EMISSIONS

by

M.S. Jennings, N.E. Krohn, R.S. Berry,
M.A. Palazzolo, R.M. Parks, K.K. Fidler

Radian Corporation
Research Triangle Park, North Carolina

np

NOYES PUBLICATIONS
Park Ridge, New Jersey, U.S.A.

Copyright © 1985 by Noyes Publications
Library of Congress Catalog Card Number 85-4923
ISBN: 0-8155-1028-4
ISSN: 0090-516X
Printed in the United States

Published in the United States of America by
Noyes Publications
Mill Road, Park Ridge, New Jersey 07656

10 9 8 7 6 5 4 3 2 1

Library of Congress Cataloging in Publication Data
Main entry under title:

Catalytic incineration for control of volatile organic
compound emissions.

(Pollution technology review, ISSN 0090-516X ; no. 121)
Bibliography: p.
Includes index.
1. Incinerators. 2. Organic compounds--Environmental
aspects. 3. Air--Pollution--Equipment and supplies.
4. Factory and trade waste--Environmental aspects.
I. Jennings, M.S. II. Radian Corporation. III. Series.
TD796.C37 1985 628.3'7 85-4923
ISBN 0-8155-1028-4

Foreword

This book describes and evaluates the use of catalytic incinerators for the control of industrial volatile organic compound (VOC) emissions. It provides an overview of how catalytic incinerators are applied to industrial VOC emissions, and assesses the overall performance typically achieved in various applications.

Catalytic incineration is a combustion control technique in which the oxidation reaction occurs at lower temperatures than thermal incineration, with the help of a catalyst, allowing oxidation reactions to occur rapidly in the temperature range 700° to 900°F. By contrast, a range of 1300° to 1500°F is required for practical oxidation rates in thermal incinerators. Significant energy savings are possible using catalytic incineration. A major attraction of catalytic incinerators is reduced fuel costs.

The emission of VOCs from industrial plants significantly increases ambient levels of photochemical oxidants, ozone, and smog. Control of industrial VOC emissions, therefore, can result in substantial environmental benefits.

Part I of the book describes catalytic incineration as applied to the control of industrial VOC emissions and assesses the overall performance, applicability, and costs as compared to alternate VOC control technologies. Part II describes the results of case studies which gathered actual performance data through a field testing program on existing operating industrial catalytic incinerators.

The information in the book is from:

> *Control of Industrial VOC Emissions by Catalytic Incineration: Volume 1. Assessment of Catalytic Incineration and Competing Controls* by M.S. Jennings, N.E. Krohn, and R.S. Berry of Radian Corporation for the Office of Research and Development, U.S. Environmental Protection Agency, July 1984.

Control of Industrial VOC Emissions by Catalytic Incineration: Volume 2. Final Report on Catalytic Incinerator Performance at Six Industrial Sites by M.A. Palazzolo, R.M. Parks, and K.K. Fidler of Radian Corporation for the Office of Research and Development, U.S. Environmental Protection Agency, July 1984.

The table of contents is organized in such a way as to serve as a subject index and provides easy access to the information contained in the book.

Advanced composition and production methods developed by Noyes Publications are employed to bring this durably bound book to you in a minimum of time. Special techniques are used to close the gap between "manuscript" and "completed book." In order to keep the price of the book to a reasonable level, it has been partially reproduced by photo-offset directly from the original reports and the cost saving passed on to the reader. Due to this method of publishing, certain portions of the book may be less legible than desired.

NOTICE

The materials in this book were prepared as accounts of work sponsored by the U.S. Environmental Protection Agency. Publication does not signify that the contents necessarily reflect the views and policies of the contracting agency or the publisher, nor does mention of trade names or commercial products constitute endorsement or recommendation for use.

Contents and Subject Index

PART I
ASSESSMENT OF CATALYTIC INCINERATION vs COMPETING CONTROLS

PART II
CASE STUDIES

Part I

Assessment of Catalytic Incineration vs Competing Controls

The information in Part I is from *Control of Industrial VOC Emissions by Catalytic Incineration: Volume 1. Assessment of Catalytic Incineration and Competing Controls* by M.S. Jennings, N.E. Krohn and R.S. Berry of Radian Corporation for the Office of Research and Development, U.S. Environmental Protection Agency, July 1984.

1. Introduction

This report describes and evaluates the use of catalytic incinerators for control of industrial volatile organic compound (VOC) emissions. The objective is to provide an overview of how catalytic incinerators are applied to industrial VOC emissions and to assess the overall performance typically achieved in various applications. To accomplish this objective the following elements are included:

- a review of current and developing catalytic incinerator technology
- a summary of present applications
- a technical and cost comparison with alternative VOC control technologies
- an analysis of performance testing methods and test data needs.

This report is part of an overall effort by the U. S. Environmental Protection Agency to evaluate currently available methods of industrial VOC control. Another part of this effort involved conducting a field testing program to assess and document the actual performance of a number of catalytic incinerators. The incinerators that were tested were applied in a variety of industries for VOC control.

1.1 SCOPE OF WORK

This report addresses industrial VOC emissions controlled by catalytic incinerators. Various definitions for the terms "industrial", "VOC", and "catalytic incinerator" have been used in the previous reports and literature. The following discussion outlines the definitions used for this study.

Industrial emissions are defined as plant emissions from storage, handling, and process equipment used in the production of both bulk and finished products. This includes nearly all stationary emission sources located in manufacturing plants. Specifically excluded, however, are emissions from the transportation sector and emissions from the combustion of fuels in boilers and heaters.

Volatile organic compounds (VOC's) are defined as organic compounds whose vapors may react photochemically in the atmosphere to form ozone or what is popularly called smog. Ozone is considered to adversely affect health and EPA has established National Ambient Air Quality Standards (NAAQS) for photochemical oxidants. Many areas routinely exceed the NAAQS for oxidants. Since VOC's contribute to the ambient oxidant levels, the reduction of VOC emissions can provide significant improvements in ambient air quality.

A catalytic incinerator is defined as a catalytic waste gas treatment device installed for the primary purpose of destructing VOC compounds. The emphasis in this report is to examine incinerators used primarily to reduce emissions of non-hazardous VOC species; control of hazardous waste gases is more specialized and involves health and safety concerns beyond the scope of this project. Also excluded from consideration are catalytic incinerators used primarily to control low concentration, malodorous emissions. This application typically requires control devices which can achieve extremely low outlet concentrations (on the order of 1 ppm or less). A final exclusion from the major discussions of this report are catalytic incinerators which are an integral part of a production process. For example, several chemical processes incorporate internal catalytic oxidizers which provide energy to drive turbines used in the processes. As another example, magnet wire enameling ovens are now being constructed with integral internal catalyst beds used to generate heat in the oven. This report concentrates on those applications where the catalytic incinerator is an "add-on" VOC waste gas treatment device, and other VOC control technologies could potentially be used as well.

1.2 APPROACH

This report examines the <u>overall</u> performance of catalytic incinerators in comparison to other technological alternatives. In order to provide a thorough and comprehensive evaluation of all technologies on a common basis, a number of evaluation criteria were developed. These criteria are used throughout the report and include:

- Environmental and Energy Considerations
- Developmental Status
- Applicability
- Operation and Maintenance Requirements
- Cost Considerations

Using these criteria, a wide variety of information sources were examined in order to develop a data base of information. These included:

- previous studies conducted by EPA and others
- vendors of catalytic incinerators and other VOC control systems
- industrial users of catalytic incinerators
- federal, regional, state, and local air pollution agencies
- air pollution control trade associations.
- industrial trade associations

An emphasis was placed on obtaining recent data from actual users of catalytic incinerators. Unfortunately, industrial users are reluctant to release information which might be sensitive and/or confidential. Contacts with air pollution control agencies yielded limited success. In many cases, agency officials lack the resources and manpower to retrieve information.

1.3 SUMMARY

1.3.1 Technology Description

A catalytic incinerator is essentially a thermal incinerator (sometimes called an afterburner) which employs a catalyst to promote the oxidation of VOC's to carbon dioxide and water. The catalyst allows the oxidation to occur at a lower temperature than would be possible without the catalyst. This lowering of required temperatures gives catalytic incinerators the potential to operate with less supplemental fuel than thermal incinerators. Supplemental fuel may be used to heat the incoming waste gas to a temperature suitable for catalytic oxidation.

Major factors affecting the performance of a catalytic incinerator include: (1) operating temperature, (2) space velocity (reciprocal of residence time), (3) VOC species and concentration, (4) catalyst characteristics, (4) presence of waste contaminants, and (5) heat recovery used. All these factors and others must be collectively considered in order to achieve optimum performance in a given application. Since the potential for reduced fuel costs is a major attraction of catalytic incinerators, the design and operation of the heat recovery system (if used) is very important. Many catalytic incinerators operate with little or no supplemental fuel. In some cases, a net overall energy savings has been realized with the installation of a recycle type heat recovery catalytic incinerator on a formerly uncontrolled process.

Catalytic incinerators commonly achieve destruction efficiencies in the range of 70 to 90 percent. For many sources efficiencies of 98 percent and higher are feasible, however, depending on the inlet concentration they may require large volumes of catalyst and high operating temperatures.[1]

1.3.2 Availability

Catalytic incinerator systems and/or components are available off-the-shelf from at least nine vendors. Each vendor tends to specialize in a particular design configuration, however, most designs are suitable for a wide range of applications. The availability of catalytic incinerators is not expected to constrain their use.

1.3.3 Developing Technology

Catalytic is a mature technology. Refinements in design are continually occurring; however, no revolutionary advances in the technology are presently identified.

1.3.4 Current Applications

Catalytic incinerators are widely used with an estimated 500 to 2,000 currently operating. Control of VOC vapors emitted by dryers and ovens used in surface coating processes are the most common applications. Other applications include organic chemical manufacturing processes, fuel storage emissions, and refinery emissions. Solvent evaporation ovens and dryers emissions are characterized by high volumes of waste gases with VOC concentrations of less than 25 percent of the lower explosive limit (LEL).

1.3.5 Alternative Technology Comparison

Carbon adsorption and thermal incineration are directly competitive VOC control alternatives in most applications. Carbon adsorption has the advantage of recovering the VOC either for reuse or as a fuel for boilers or heaters. However, carbon adsorption applications are limited with respect to waste gas temperature (less than 100°F) and composition (carbon beds are susceptible to masking and fouling). Another constraint is the size and relative complexity of carbon adsorption systems. In cases where the VOC concentration is 10 percent LEL or higher and the recovered solvent is valuable (greater than $0.75 gal), carbon adsorption, if feasible, is usually the lowest cost alternative.

Thermal incineration is similar in emission control performance to catalytic incineration. Capital costs are also similar; however, the reduced fuel requirements of catalytic incinerators causes total annualized costs to be lower in most cases. Three exceptions are noted where thermal incineration may be preferred over catalytic incineration: (1) For exhaust streams which contain significant amounts of catalyst poisons and/or fouling agents, thermal incinerators may be the only technically feasible control. (2) Where extremely high VOC destruction efficiencies of difficult to control VOC species are required, thermal incinerators may attain higher performance. (3) For relatively rich VOC waste gas streams (20-25 percent LEL and higher) the potential for catalyst overheating may require the addition of dilution air to the waste gas stream. This increases the costs of catalytic incinerator control and makes thermal incineration cost competitive.

A relatively new (first installed in 1976) thermal incinerator system using regenerative heat exchange is now available. This system is reported to use significantly less fuel than thermal incinerators with conventional recuperative heat exchangers. Depending on overall costs, this reduction in fuel costs could make thermal incinerators cost competitive at LEL levels of less than 20 percent. However, insufficient data were available to conduct direct cost comparisons for this study.

1.3.6 Performance Measurement Methods for Field Testing

To assess the overall performance of catalytic incinerators, both process/incinerator characterization and emission measurements are required. Process/incinerator measurements collected would include temperatures, pressures, flow rates, and fuel use. Emission measurements would include VOC concentrations and gas flow rates in the inlet and outlet of the incinerator.

A variety of VOC measurement methods are available. To insure that data would be collected using standardized and accepted procedures, EPA reference methods are recommended. A combination of Method 25A and draft Method 18 provides several advantages. Method 25A, utilizing a flame ionization analyzer (FIA) could be run continuously over several days to allow the effect of process fluctuations to be assessed. A disadvantage of Method 25A is the lack of compound specific VOC concentration information (the FIA gives only a relative concentration with respect to a calibration gas). Method 18 involves gas chromatography analysis which provides compound specific concentrations. An advantage of using both simultaneously is that the Method 18 results could be used to obtain specific destruction efficiencies while Method 25A results could be used to get total hydrocarbon destruction efficiencies.

1.3.7 Testing Needs

The current information data base concerning catalytic incinerator performance is deficient in several respects. What data are available have been collected with varying test procedures and are lacking in documentation. Most emission data were collected with the sole objective of determining compliance with emission standards. For this reason, other factors pertinent to catalytic incineration performance (such as fuel use) were not measured and/or not recorded.

This study recommends that a comprehensive testing program be initiated to test in the field a variety of operating catalytic incinerators. The test program would provide an assessment of the overall performance of catalytic incinerators in a variety of applications. Test sites would be selected to cover a range of applications and waste gas characteristics.

REFERENCE

1. Rolke, R. W., et al. Afterburner Systems Study. (Prepared for U. S. Environmental Protection Agency.) Research Triangle Park, N.C. Publication No. EPA-R2-72-062. August 1972.

2. Evaluation Criteria

This chapter provides a brief review of the criteria used to evaluate catalytic incineration and other VOC control techniques. The criteria presented allow evaluation on a common basis. The criteria were chosen so as to simulate the evaluation process which might be used to select a VOC control for a given application.

The primary evaluation criteria used in this report fall into the following areas:

- Environmental and Energy Considerations
- Developmental Status
- Applicability
- Operation and Maintenance Requirements
- Cost Considerations

2.1 ENVIRONMENTAL AND ENERGY CONSIDERATIONS

This criteria category is designed to evaluate the potential of each control technology for reducing VOC emissions without major undesirable secondary pollution problems. In addition, the energy requirements of the control technique are examined. The considerations evaluated include:

1. Potential Overall VOC Removal Efficiency - What is the overall VOC removal potential of the control technique? Does the efficiency vary with specific compounds?

2. VOC Removal Variability - How does the performance degrade with time and with process variations?

3. Secondary Air Pollution Generation - Does the technology increase emissions of undesirable secondary pollutants?

4. Solid Waste Pollution Generation - Does the technology increase amounts of solid waste which must be disposed?

5. Water Pollution Generation - Does the technology increase amounts of waste water which must be disposed?

6. Supplemental Energy Use - What are the supplementary energy requirements of the technology? Does the technology offer the possibility of significant heat recovery back to the process?

In applications which require relatively high VOC removal efficiencies, the overall VOC removal potential of the technology will be the overriding consideration. However, in many instances VOC removal requirements are well below the potential of the technology. Incinerators can achieve very high removal efficiencies (greater than 98 percent) in most applications, however, they must be operated at very high temperatures and designed with large catalyst volumns. Usually they are not operated in this manner in order to reduce supplemental fuel and catalyst requirements.

2.2 DEVELOPMENTAL STATUS

This category is designed to evaluate the availability of the control technique to potential users. The following items are addressed:

1. Overall Developmental Status - What is the present status of development? Is significant R&D still required for some applications?

2. Commercialization - Are vendors presently able to supply units in reasonable lengths of time?

2.3 APPLICABILITY

This category examines the applicability of the control technique to meeting various emission control requirements. This examination looks at the following considerations:

1. Waste Gas Temperature Requirements - What are the temperature requirements for the waste gas stream before the technology can be used?

2. Susceptibility to Contaminants - What level of contaminants is acceptable in the waste gas?

3. Sensitivity to VOC Species - Does the performance degrade with certain VOC species or mixtures of species?

4. Sensitivity to Varying VOC Concentrations - Does the performance degrade when the VOC loading to the device changes?

5. Size and Space Requirements - Does the technology impose size and space constraints with respect to retrofits?

6. Emission Data Requirements - Does the technology require extensive emissions tests or pilot plant operations to insure successful application?

2.4 OPERATION & MAINTENANCE (O&M) REQUIREMENTS

This category is designed to evaluate the overall O&M required to operate a control system. O&M considerations include the following:

1. Overall Equipment Life - What useful life can be expected before replacement is required?

2. Overall reliability - Does the system present any unusual design features which suggest frequent breakdowns? What is the experience of industry concerning reliability?

3. Operation Labor Requirements - What level of operator involvement is required?

4. Maintenance Requirements - What level of maintenance is required and what are the items which require maintenance?

5. Startup/Shutdown Procedures - Are the procedures complex and do they constrain the use and operation of the control device?

6. Potential for Equipment Damage - Does improper operation create the potential for unit damage? Does the system present the possibility of a safety hazard with improper operation?

2.5 COST CONSIDERATIONS

In most instances, costs will be the primary decision tool in selecting a control technique. Two types of costs are used in the economic evaluation:

1. Capital Investment - What is the total capital required to build and make operational the control system?

2. Annualized Costs - What is the total annual (capital charges plus operating) cost required to build, operate, and maintain the control system?

For small companies where capital may be limited, capital costs may be the most important cration. For larger concerns where capital is not a constraint, annual costs are usually of primary importance.

3. Review of Present Status of Catalytic Incineration Technology

This chapter presents a review of the present status of catalytic incineration technology. Using the evaluation criteria presented in Chapter 2 as a basis, this chapter provides a broad overview of how catalytic incineration is applied in the control of industrial volatile organic compound (VOC) emissions.

As discussed in Chapter 1, this study is limited to catalytic incinerators used for industrial VOC control. Excluded are the following catalytic devices:

- Odor control incinerators
- Catalytic oxidizers used as energy recovery devices integral to chemicals production
- Catalytic devices in which the catalyst is internal to the process
- Non-industrial VOC controls (i.e. mobile sources)
- Hazardous waste incinerators

3.1 STATUS OF THE TECHNOLOGY IN USE

Catalytic incinerators are devices used to control combustible vapor emissions in gas streams. Contact of a VOC containing waste gas stream with the catalyst allows oxidation of the VOC to occur within a short time period. The catalyst, without being altered, accelerates the oxidation reaction by concentrating the reactants (VOC and oxygen) at the active sites on the catalyst. This greatly accelerates the reaction rate and allows the reaction to occur at lower temperatures.[1,2]

This section addresses three areas of catalytic incineration technology:

- Operation and Maintenance Characteristics
- Performance
- Availability

The section opens with a brief description of the basic elements of a catalytic incinerator followed by a discussion of the factors affecting the incinerator's performance.

3.1.1 Description of the Catalytic Incineration System

The basic elements of a catalytic incinerator include the following:

- preheat burner
- mixing chamber
- catalyst bed
- heat recovery equipment

An example of the layout of a typical catalytic incinerator is shown in Figure 3-1.

In order to oxidize the VOC's, the waste gas stream must be heated to a minimum temperature. The temperature varies depending on the VOC compound and concentration. A temperature of 600°F will generally insure adequate VOC destruction, however in many installations a lower temperature is adequate to achieve the required destruction. The preheat burner is used to heat the waste gas stream to the required temperature. Many catalytic incinerators operate with natural gas-fired burners without a forced draft air supply (the waste gas stream supplies air for combustion). Catalytic incinerators may also be oil-fired.

The waste gas should be of uniform temperature before it enters the catalyst bed. The mixing chamber, downstream of the preheat burner, distributes the hot, combustion products from the burner into the waste gas. Proper mixing insures a uniform waste gas temperature prior to entering the catalyst bed.

Following the mixing chamber is the catalyst bed. The catalyst bed usually consists of finely divided platinum or other metal deposited on a ceramic or metal support structure. The finely divided metal is the actual catalyst for the oxidation reaction.[3]

The catalyst bed may consist of a fixed (single) bed, a modularized bed or a fluidized bed. A fixed bed, depicted in Figure 3-1, is assembled from precut pieces or pellets that are packed within a single vessel. A modularized bed consists of a number of modules or cartridges of catalyst placed within a single vessel. A fluidized bed consists of spherical pellets supported by a grate or screen in a single vessel.[1,4] Air movement through the bed causes the pellets to move about at random and behave as a fluid.

Although not essential for operation, heat recovery equipment is usually included as part of the incineration system. The heat produced from the combustion of the VOC's in the catalyst bed, along with the heat

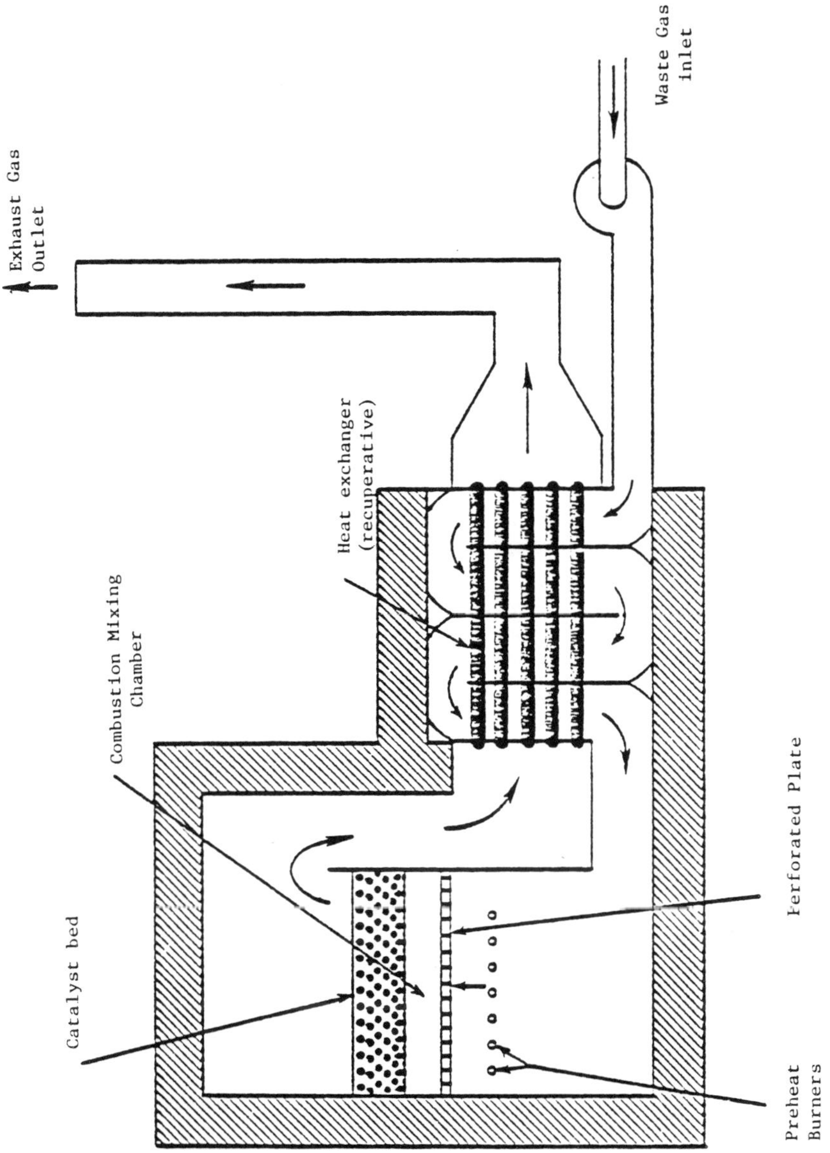

Figure 3-1. Catalytic incinerator with primary heat recovery.

contained in the waste gas prior to entering the bed, can be recovered for use. The recovered heat can be used in several ways:

- to heat the waste gas and reduce the energy use of the preheat burner,
- to heat the process itself by recycling the exhaust back to the process
- to produce steam or hot water in a waste heat boiler.

The recuperative heat exchanger shown in Figure 3-1 transfers heat from the hot, exhaust gas to the colder, inlet waste gas.

Figure 3-2 presents a number of possible heat recovery configurations using recuperative, recycle, or both types of heat exchange. Many schemes are possible; this figure identifies a few common schemes. Configurations 1 or 3 would be well suited to use with a waste heat boiler since the incinerator exhaust gases are relatively hot compared to the cooler exhaust gases exiting the heat exchangers in Configurations 2, 4, and 5. The optimum configuration in terms of overall cost and energy use will depend on a wide variety of site specific factors.

3.1.2 Factors Affecting Performance

There are three major criteria which are used to evaluate the performance of a catalytic incinerator:

- VOC destruction efficiency
- VOC concentration in the exhaust gas
- Energy use.

In many applications, well designed, well operated catalytic incinerators are capable of achieving high VOC destruction efficiencies, low VOC concentration in the exhaust gas and low energy usage.

The performance of a catalytic incinerator is strongly affected by several factors.[2,5] These factors are listed in Table 3-1. These factors are interrelated through the mechanism of the catalytic oxidation reaction and must be considered collectively when designing a catalytic incineration system.

Operating temperature is a major factor affecting the VOC destruction efficiency of a catalytic incinerator. The operating temperature needed to achieve a particular VOC destruction efficiency depends primarily on the species of pollutants contained in the waste stream, the concentration of the pollutants, and the catalyst type. Generally, higher catalyst operating

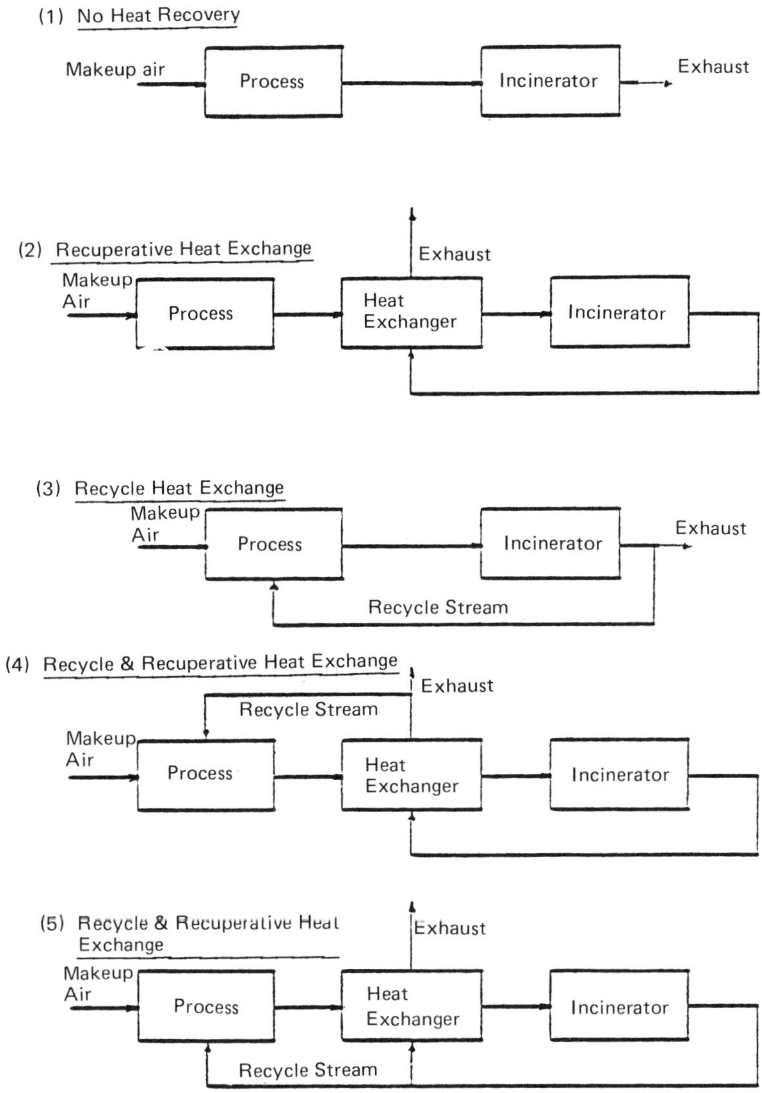

Figure 3-2. Heat recovery configurations.

TABLE 3-1. FACTORS AFFECTING CATALYTIC INCINERATOR PERFORMANCE

Operating Temperature
Space Velocity
VOC Concentration & Species
Catalyst Characteristics
Presence of Poisons or Masking Agents in Stream
Heat Recovery

TABLE 3-2. TEMPERATURES FOR 90% CONVERSION IN A CATALYTIC INCINERATOR[6]

Component	Temperature - °F
Hydrogen	105
Acetylene	200
Carbon Monoxide	220
Propyne	240
Propadiene	250
Propylene	260
Ethylene	290
n-heptane	300
Benzene	300
Toluene	300
Xylene	300
Ethanol	315
Methyl ethyl ketone	370
Methyl isobutyl ketone	370
Propane	410
Ethyl acetone	415
Dimethyl formamide	425
Ethane	430
Cyclopropane	455
Methane	490

temperatures increase the VOC destruction efficiency; however, they also increase the energy usage. Each pollutant has a temperature which must be reached to initiate the catalytic oxidation reaction. The initiation temperature is also dependent on the type of catalyst used in the incinerator. Table 3-2 lists initiation temperatures for several compounds when in contact with a precious metal catalyst as reported by an incinerator vendor. The temperature required to provide a given destruction efficiency will vary depending upon the specific catalyst, VOC concentration, and residence time in the incinerator. Table 3-2 is intended to provide a comparison between temperatures required for various VOC species.

The VOC destruction efficiency is also strongly influenced by the space velocity of the gas through the catalyst bed. Space velocity is commonly defined as the volumetric flow of gas entering the catalyst bed chamber divided by the volume of the catalyst bed. As such, it is the reciprocal of residence time for the waste gas in the catalyst bed. A space velocity of 50,000 hr^{-1} is equivalent to a contact time of 0.072 seconds.[7] The relationship between space velocity and VOC destruction efficiency is strongly influenced by the catalyst operating temperature. This relationship is illustrated in Figure 3-3. As space velocity increases, destruction effi- ciency decreases, and as temperature increases, VOC destruction efficiency increases.

As the VOC concentration of the waste gas increases, the temperature difference between the inlet and outlet of the catalyst bed chamber increases due to the increased VOC combustion. With the increased VOC combustion and higher outlet temperatures, more preheat can be supplied through heat recovery of the exhaust gases rather than from the preheat burner.[10] With many systems, a point is reached where the incinerator will achieve satisfactory destruction efficiency without supplemental fuel requirements. The point is dependent on the inlet waste gas stream temperature and VOC concentration, and the amount of heat recovery equipment used in the system. Higher VOC waste gas concentrations increase the energy efficiency of the catalytic incineration system, however, insurance and plant safety regulations usually require waste gas concentrations to be below 25 percent of the Lower Explosive Limit (LEL). Even with the 25 percent LEL constraint, low energy usage and high VOC destruction efficiencies are achievable. Table 3-3 lists the Lower Explosive Limits for several compounds. In streams containing multiple VOC compounds, it is common to set the LEL of the most explosive compound as the LEL of the mixture.[9]

The configuration and type of catalyst are important factors affecting VOC destruction efficiency. The catalyst bed consists of a metal or metal oxide catalyst deposited on a metal or ceramic catalyst support. Precious metals, typically platinum or palladium, are usually preferred over nonprecious metals because precious metals generally allow lower operating temperatures and higher space velocities for a given destruction efficiency.[5]

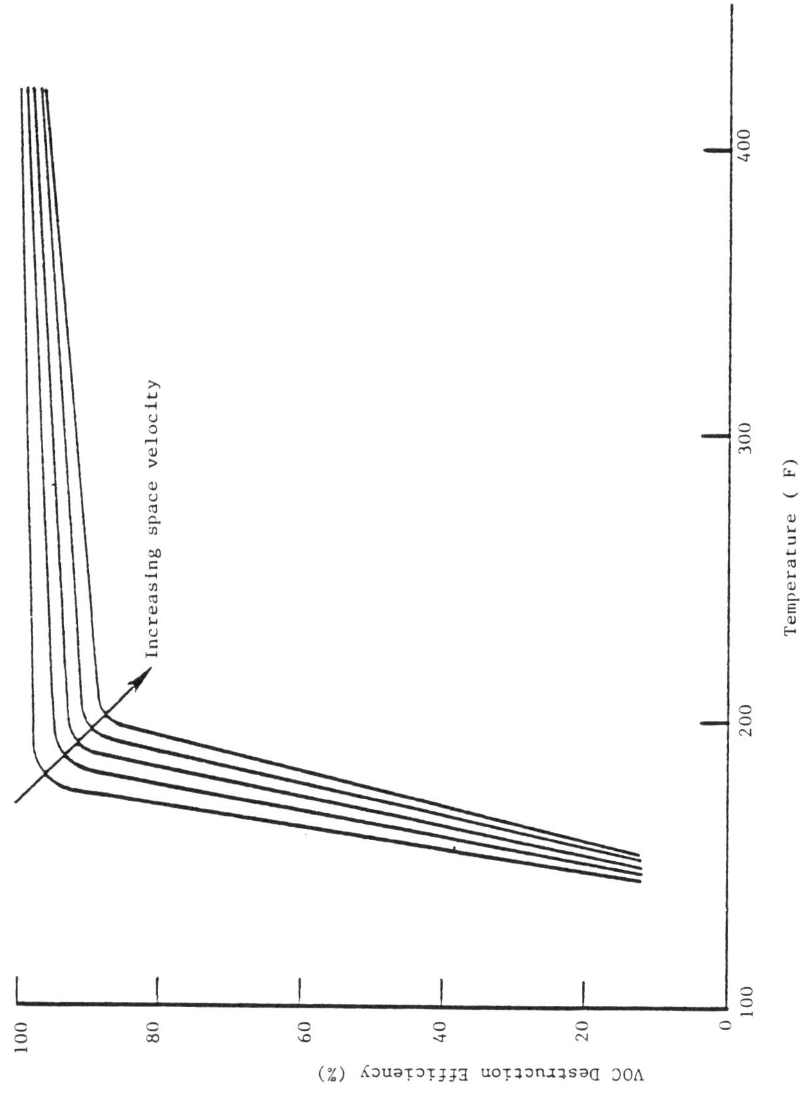

Figure 3-3. Combined effects of catalyst operating temperature and space velocity on conversion.8

TABLE 3-3. LOWER EXPLOSIVE LIMITS[6]

Contaminant	LEL-%Vol.
Carbon Monoxide	12.5
Hydrogen	4.0
Butane	1.9
Ethane	3.0
Ethylene	3.1
Propylene	2.4
Styrene	1.1
Benzene	1.4
Xylene	1.0
Toulene	1.4
Methyl Alcohol	7.3
Isopropyl Alcohol	2.0
Acetone	3.0
Methyl Ethyl Ketone	1.8
Methyl Acetate	3.1
Cellosolve Acetate	1.7
Acrolein	2.8
Cyclohexanone	1.1
Acetaldehyde	4.0
Furfural	2.1

Metal oxide ceramics such as mullite (aluminum silicate), alpha-alumina, cordierite, and spinel (magnesium aluminate) are the major materials used for ceramic supports. Nickel alloys are sometimes used for metal supports because they are easy to fabricate and they resist oxidation.[10] Stainless steel supports are also used.[10] In using either a ceramic or metal support, a textured support surface is preferred over a smooth surface because the textured surface enhances catalyst adhesion.[1]

The catalyst support and bed geometry are critical factors in relation to the economic and operational characteristics of the catalytic incineration system. The catalyst support and bed geometry influence the size and shape of the catalyst bed chamber which in turn affects the cost, and the pressure drop across the bed (which in turn affects blower power consumption).[5] Table 3-4 contains a listing of conventional bed geometries.

The size of the catalyst bed in the incinerator depends mainly on the degree of VOC reduction required. VOC destruction efficiencies up to 95 percent can usually be attained with reasonable space velocities. However, the catalyst volumes required for extremely high conversions (e.g. 99%) can sometimes make catalytic incineration uneconomical for specialized odor control applications where extremely high conversions are required.[3]

The specific type of catalyst used in the incinerator depends on the type of compounds being oxidized. Different VOC compounds require different catalysts. Knowledge of the composition of the VOC waste gas stream is necessary in order to design the proper catalyst into the catalytic incineration system.

The presence of particulates or catalyst poisons in the waste gas stream can have a severe effect on the catalyst and all three performance criteria. Particulates and poisons reduce the activity of the catalyst for promoting the oxidation reaction. Particulates mask and coat the catalyst preventing contact between the VOC's and the catalyst's surface. Poisons combine with the catalyst and thus change the catalyst's activity.

The effects of poisons or particulates can sometimes be overcome by increasing the catalyst operating temperature. Periodic temperature elevations will sometimes burn off the poisons or particulates. Also, the catalyst can be removed and washed to remove these substances. However, a catalyst which is irreversibly poisoned must be replaced.

Low levels of particulates or potential poisons can sometimes be tolerated without a dramatic decrease in performance. Generally it is recommended that the maximum particulate concentration not exceed 0.05 grains/scf and that the maximum poison concentration not exceed 25 ppm.[11] Table 3-5 lists several poisons and masking substances and their effects.

TABLE 3-4. CONVENTIONAL BED GEOMETRIES[1,3,8,10]

Geometry	Advantages	Disadvantages
Metal ribbons	Low pressure drop; high surface to volume	Less active than ceramic supported catalysts
Spherical pellets	Can be used in both fixed and fluidized beds	High pressure drop; attrition problem
Ceramic rods	Low pressure drop	Low surface area to volume ratio
Ceramic honeycomb	Low pressure drop; high surface area to volume ratio	May have non-uniform catalyst coating
Metal honeycomb	Low pressure drop; high surface area to volume ratio; high mechanical strength	Less active than ceramic supported catalysts

TABLE 3-5. SUBSTANCES THAT INFLUENCE CATALYST ACTIVITY[2,3]

Type of Inhibitor	Effect
Fast Acting Inhibitors	
Phosphorus, Bismuth, Lead, Arsenic, Antimony, Mercury	Irreversible reduction of catalyst activity at rate dependent on concentration and temperature.
Slow Acting Inhibitors	
Iron, Tin, Silicon	Irreversible reduction of catalyst activity. Higher concentrations than those of fast acting catalyst inhibitors may be tolerated.
Reversible Inhibitors	
Sulfur, Halogens, Zinc	Reversible surface coating of catalyst active area at rate dependent on concentration and temperature.
Surface Maskers	
Organic Solids	Reversible surface coating of catalyst active area. Removed by increasing catalyst temperature.
Surface Eroders and Maskers	
Inert Particulate	Surface coating of catalyst active area. Also errosion of catalyst surface at rate dependent on particle size, grain loading and gas stream velocity.

The use of heat recovery equipment in the incineration system can reduce the energy usage of the incinerator. Heat recovery may be accomplished through recuperative heat exchange, regenerative heat exchange, recycle heat exchange, or waste heat boilers. Recuperative heat exchangers direct the high temperature and low temperature streams through channels that keep them separate yet provide a surface for heat transfer. The conventional shell and tube heat exchanger is the most common type of recuperative heat exchanger. The temperature difference between the hot exhaust gas entering the exchanger and the cool waste gas entering the exchanger establishes the maximum amount of heat which can be transferred. The waste gas temperature rise can never exceed this difference and the degree to which this theoretical limit is approached measures the heat transfer effectiveness. The heat transfer effectiveness factor is defined as follows:

$$E = \frac{Q}{Q_{max}} = \frac{C_{p-wg}\,(T_{wg-out} - T_{wg-in})}{C_{p-wg}\,(T_{eg-in} - T_{wg-in})}$$

where:

E = heat transfer effectiveness (fraction)

Q = heat transferred (Btu/lb)

Q_{max} = maximum potential heat transfer Btu/lb)

C_{p-wg} = specific heat capacity of the waste gas (Btu/lb-°F)

T_{wg-in} = temperature of cool waste gas entering exchanger (°F)

T_{wg-out} = temperature of hot waste gas leaving exchanger (°F)

T_{eg-in} = temperature of hot exhaust/flue gas entering exchanger (°F)

Heat transfer effectiveness factors up to 85 percent are achievable with recuperative exchangers applied to both catalytic and thermal incinerators,[93] however, factors above 65 percent are usually not economical with recuperative exchangers.

A regenerative heat exchanger temporarily stores energy from a hot stream in a heat sink such as a ceramic bed. The hot stream is directed through the ceramic bed until the temperature of the bed nearly reaches that of the hot stream. Direction of streams are changed and a cold gas stream is passed through the hot, ceramic bed. The cold stream is thus heated by the hot, ceramic bed. When the bed cools, directions of streams are reversed again, repeating the cycle. Heat recovery effectiveness factors up to 85 percent and higher are achievable with regenerative heat exchangers.[12,13,14] To date, regenerative exchangers have been applied to thermal incinerators only.

Heat recovery using recycle involves directing a portion of the hot exhaust stream to some other process or operation where heat is needed, typically the process emitting the VOC. Typically about 30 percent of the exhaust gas is recycled to the process.[3]

Heat recovery can greatly increase the energy efficiency of the incineration process. However, the economics and practicality of heat recovery should be weighed carefully when designing the catalytic incineration system. The following factors should be considered.[3]

1. Feasibility - There must be a need or use for the recovered heat. Some potential uses are listed in Table 3-6.

2. Safety - Excessive preheat of streams containing combustible material might lead to flammable or even explosive conditions. Recycle of a hot exhaust stream could introduce contamination hazards in process ovens.

3. Dependability - If the incinerator is an intermittent or undependable source of heat, the heat recovery system will also operate intermittently. If the intermittent operation of the heat recovery system in turn shuts down the process unit, the economic value of the energy savings could be dramatically reduced.

4. Costs - A cost evaluation should be performed to determine the amount of heat recovery that is economically justifiable. An economic evaluation must involve a trade off of the capital required to build the recovery system vs. the fuel savings.

5. Available Space - Heat exchange equipment is often large and requires large ductwork. Space considerations must be taken into account especially in retrofit situations.

6. Possible Bypass Problems - A leaky exchanger can allow pollutants to bypass the incinerator, reducing the destruction efficiency of the incinerator.

3.1.3 Operation and Maintenance Characteristics

Operation of a catalytic incinerator is relatively simple and maintenance is relatively low. Operation and maintenance requirements are generally similar to requirements for thermal incinerators and less than requirements for other common VOC controls. The major operating problem associated with catalytic incinerators is catalyst deactivation.

TABLE 3-6. POTENTIAL USES FOR HEAT RECOVERED FROM THE CATALYTIC INCINERATOR[2,13]

Use	Conventional Method
Preheat inlet stream to incinerator	Recuperative or Regenerative
Heat curing oven or dryer	Recycle
Generate steam	Waste heat boiler
Generate hot water	Waste heat boiler
Evaporate water or solvent in other processes	Recycle

Deactivation may be the sole result of physical changes in the catalyst such as recrystallization of the catalyst support or recrystallization of the catalyst metal. These types of deactivation are the result of constant exposure to heat and thermal effects and are referred to as thermal aging.[1]

Catalyst deactivation may also be the result of an accumulation of contaminants that form a barrier or mask between the waste gas stream and catalyst. Typical surface masking problems are caused by dust or dirt in the waste gas stream, metal oxides formed from materials in the process (e.g. silica from silicones), corrosion products from the duct system or organic char formation caused by low reactor operating temperatures. Combustible masking materials such as organic char may be partially or completely removed by periodic elevations of the catalyst bed temperature. Non-combustible masking materials may be removed by air lancing or aqueous washings.[15]

Catalyst deactivation can result from a chemical change in the catalyst caused by the combination of thermal effects and contamination. This deactivation or poisoning reduces the catalyst's activity and is caused by a reaction between the contaminant and the catalyst metal. True catalyst poisoning is infrequently encountered in typical catalytic incineration systems, but when it is encountered, its effect on the catalyst is irreversible.[1,3]

Certain process accommodations may be used to decrease the risk of catalyst deactivation.[3]

1. Avoid using galvanized metal in process ovens or ductwork since zinc is a catalyst poison. Stainless steel, aluminized steel and carbon steel are usually satisfactory replacements.

2. Avoid placing mercury glass thermometers in locations where breakage could contaminate the waste stream.

3. Avoid using phosphate cleaning detergents around the process since phosphorus is a catalyst poison.

4. Avoid using lubricants containing silicones or inorganic metal compounds which might poison the catalyst upon evaporation.

A catalyst's lifetime typically ranges between 2 and 5 years.[1,4,5,7] Deactivation due to thermal aging occurs gradually over the lifetime of the catalyst. A catalyst's lifetime may be temporarily extended by increasing the catalyst bed temperature. Catalyst bed temperature can be increased by raising the temperature of the waste gas just prior to entering the catalyst bed. The higher preheat temperature increases the reactivity of the reactants (VOC and oxygen) thereby temporarily overcoming a loss of catalyst activity. This practice of raising the preheat temperature is normally followed until the added fuel cost for higher preheat makes catalyst replacement a more economically attractive choice. The practice of raising preheat temperature may also be discontinued if the catalyst bed outlet temperature has reached a point where catalyst activity is rapidly declining.[3]

No special operator training is required for normal operation of a catalytic incineration system. Typical O&M requirements are 0.5 manhours/shift.[54] Most units are automatically controlled and require no special operator attention. Some care must be exercised with catalytic incinerators especially during start-up, shut-down and storage, however. Also, catalyst beds must be kept clean and relatively dry in order to maintain their activity. The degree of sensitivity and care required will depend on the type of catalyst. Specific care and maintenance procedures are typically provided by the manufacturer.[15]

The primary operating control for a catalytic incinerator is typically the outlet temperature (downstream from the catalyst bed) which is used to adjust the fuel supply to the preheat burner. Most manufacturers provide a thermocouple to sense this temperature and an indicating (or recording) proportional controller to actuate the fuel valve. A second thermocouple in this downstream location and a third thermocouple upstream of the catalyst bed (which senses the temperature of the preheated stream) are usually connected to a temperature indicator (or recorder) and high temperature limit instrument. These function to prevent overheating the catalyst or incinerator system in the event of a fuel controller failure or a sudden dramatic increase in the VOC concentration of the waste gas stream. The temperature difference across the catalyst bed gives an indication of the activity of the catalyst and provides a basis for (manually) resetting the outlet (control) temperature or for making decisions regarding cleaning or replacement of catalyst.[3]

Safety instrumentation such as LEL meters and alarms are normally included in a catalytic incineration installation. A safety bypass around the primary heat recovery system is also normally included to allow the waste heat gas to bypass the heat exchanger to prevent system overheating.[16]

3.1.4 Performance Characteristics

In order to assess the present status of catalytic incinerators, quantitative performance data on catalytic incinerators currently in use were needed. Several avenues were pursued in attempting to obtain this data. All ten Environmental Protection Agency Regional Offices were contacted. All offices suggested contacting state agencies. Letters were sent to eight state and local agencies requesting permits and test reports containing documented performance information on catalytic incinerators used for VOC control. The agencies contacted included the following:

- Georgia Environmental Protection Division,

- Illinois Environmental Protection Agency,
- Indiana State Board of Health,
- Massachusetts Division of Air Quality and Hazardous Materials,
- New Jersey Bureau of Air Quality Control,
- Ohio Environmental Protection Agency,
- Pennsylvania Bureau of Air Quality Control,
- Texas Air Control Board, and
- South Coast Air Quality Management District.

These particular agencies were contacted because plants using catalytic incinerators had been located in their areas of jurisdiction. Several emission test reports were obtained as a result of this effort.

A literature search was conducted to locate any previously reported performance data. In addition to finding reported test data, plant trip reports and plants using catalytic incinerators were located during this effort.

Tables 3-7 and 3-8 contain a summary of the catalytic incinerator performance data located during this study. Table 3-7 is a summary of data obtained from emission test reports and Table 3-8 is a summary of previously reported data. Because plant names and locations were not mentioned in the previously reported data, it is not known whether the two tables contain duplicated information. Therefore, only Table 3-7 will be used as representative of catalytic incinerator performance. The performance data may be summarized as follows:

- VOC destruction efficiencies ranged from 42.5 to 99 percent
- VOC exhaust gas concentrations ranged from 15 ppm to 818 ppm
- Energy usage was not usually reported.

As can be seen from the data, the highest efficiency, 99 percent was attained when the waste gas VOC concentration was very high at 21000 ppm. Generally these very high destruction efficiencies are achievable only with relatively rich VOC streams.

No data were located concerning the lowest VOC exhaust gas concentration achievable with catalytic incineration. The test data show a wide variation in exhaust gas concentrations. Variations in VOC destruction efficiencies were less; VOC destruction efficiencies averaged 95.6 percent.

Table 3-7. Summary of Catalytic Incinerator Performance Data

Plant	Process	Test Details			Operating Temperature (°F)	Heat Recovery	Flow Rates		Concentrations		Efficiency (percent)	Energy Usage (Btu/hr)
		Date	Method of VOC Analysis	Length (min)			Inlet (scfm)	outlet (scfm)	Inlet (ppm)	outlet (ppm)		
Pacific Press[17] Verilon, CA	Printing press	1/24/80	TCA	30	810	--	8000	8800	728	39	94	
National Can[18] Corporation Etiwanda, CA	Drying Oven	1/14/76	Thermal conduct-ivity	15	650	yes	18892	18892	1800	160	91	
	Drying Oven	1/13/76	Thermal conduct-ivity	15	680	yes	15335	15335	2700	160	94	
	Drying Oven	2/03/77	Thermal conduct-ivity	15	700	yes	2371	2371	4200	300	93	
	Drying Oven	3/08/77	Thermal conduct-ivity	15	800	yes	1939	1939	5000	320	94	
	Drying Oven	9/18/80	TCA	30	750	yes	3800	3800	2293	122	94.7	
	Drying Oven	9/18/81	TCA	30	850	yes	3800	3800	3121	90	97.1	
	Drying Oven	9/18/81	TCA	30	650	yes	3900	3900	3045	200	91.6	
Continental[17] Can Co. Los Angeles, CA	Litho Oven	9/19/73	TCA	45	1035	yes	3320	3290	6390	818	87	2,800,000
	Litho Oven	9/19/73	TCA	45	1145	yes	3320	3290	6800	530	90	2,800,000
	Litho Oven	11/13/75	TCA	38	800	no	2600	2600	3920	346	91	4,200

Table 3-7. (Continued)

Plant	Process	Test Details: Date	Test Details: Method of VOC Analysis	Test Details: Length (min)	Operating Temperature (°F)	Heat Recovery	Flow Rates: inlet (scfm)	Flow Rates: outlet (scfm)	Concentrations: inlet (ppm)	Concentrations: outlet (ppm)	Efficiency (Percent)	Energy Usage (Btu/hr)
E.I. DuPont[19] Beaumont, TX	Acrylonitrile Absorber	9/13/77			750	yes	76115		4327[a]	2489[a]	42.5	2,187,000
American Can[28] Co. Chicago, IL	Coating Line	4/30/75	FID		730			6293	530	15	97.1	
	Coating Line	5/01/75	FID		735			5585	227	15	93.4	
	Coating Line	5/02/75	FID		720			3987	478	20	95.8	
Kaiser[21] Aluminum Toledo, OH	Paint Line	6/19/79	Carbon Tubes	30	520	no	12600 5480	23400	244[a]	19.7[a]	92.2	
Uniroyal, Inc.[22] Mishawaka, IN	5 post dry ovens	6/20/78	FID			yes	4300		2800			
	2 spreader ovens	9/19/78	FID			yes	6200		13270			
	Spreader/oven	5/12/80	FID			yes	5956	7319	1350	115	90.6	
	Formaldehyde[7] plant	11/30/79	GC		1000		1000		2000		97.9	
		12/18/79	GC		800		3000				98.5	
		1/11/80	GC		1000		1740				98.3	
		3/13/80	GC								97.9	
		3/18/80	GC		940		2400				98.3	
		3/25/80	GC		950		2700				98.0	

[a]These data reported in lb/hr.

Table 3-7. (Continued)

Plant	Process	Test Details: Date	Test Details: Method of VOC Analysis	Test Details: Length (min)	Operating Temperature (°F)	Heat Recovery	Flow Rates: Inlet (scfm)	Flow Rates: Outlet (scfm)	Concentrations: Inlet (ppm)	Concentrations: outlet (ppm)	Efficiency (percent)	Energy Usage (Btu/hr)
Western Metal[17] Decorating Co. El Segundo, CA	Litho Oven	1/13/69	TCA	20	980		1530	1410	14265	511	97	
	Litho Oven	3/04/69	TCA	30	1050		1410	1270	9390	673	94	
	Litho Oven	3/21/72	TCA	50	1030		1410	1370	21000	180	99	
Ambassador[17] College Pasadena, CA	Printing Press	12/3/74	TCA	47	1170	no	1423	1400	3835	370	90.5	6,400,000
Medallion[17] Graphics Los Angeles, CA	Lithographic Press	12/3/75	TCA	50	850	no	2190	2300	602	99	86	960,000

Notes: TCA = Total combustion analysis.
FID = Flame ionization detector.
GC = Gas chromatography.

TABLE 3-8. SUMMARY OF PREVIOUSLY REPORTED DATA FOR CATALYTIC INCINERATOR PERFORMANCE[24]

Process	Operating Temperature (°F)	Emissions Inlet (lb/day)	Emissions Outlet (lb/day)	Afterburner Efficiency (%)
Lithographic oven	580			
Lithographic oven	690			
Lithographic oven		1845	260	86
Coating oven				
Coating oven		245	71.4	71
Lithographic oven		382	304	20
Lithographic oven	702	965	470	51
Lithographic oven		1010	742	27
Litho-line oven	806	2337	573	76
Lithographic oven	637	1716	403	77
Wire coater		168	6.3	96.2
Wire coater		99.7	1.65	98.3
Lithographic oven	630	801	121	85

As can be seen from the tables, several different VOC measurement methods were used to determine VOC concentrations: Total Combustion Analysis, (TCA) is a method commonly used by the South Coast Air Quality Management District and is similar to EPA Reference Method 25. VOC measurement by thermal conducitivity is accomplished using thermal conductivity analyzers. These analyzers operate on the principle that different gases conduct heat differently. Measurement using carbon tubes requires the sample gas be withdrawn over activated charcoal. The organic materials are adsorbed on carbon and later desorbed and measured using gas chromatography. FID, Flame Ionization Detection is similar to EPA Reference Method 25A. GC, Gas Chromatography is similar to draft EPA Reference Method 18 (see Chapter 6 for further discussion of EPA Reference Methods 18, 25, and 25A).

At this time there is insufficient data to determine a quanitative relationship between test results measured by each method. However, since efficiencies are calculated by measuring the inlet and outlet with the same methods, the efficiency data in Table 3-7 should be readily comparable across the entire data base.

3.1.5 AVAILABILITY

Through contact with the Industrial Gas Cleaning Institute and a literature search, ten vendors of catalytic incineration system components were identified. These were:

- Air Resources, Inc.
- Catalytic Products, International
- CE Air Preheaters
- E.I. du Pont de Nemours & Company, Inc.*
- Engelhard
- Johnson Matthey, Inc.
- Met-Pro Corporation
- UOP, Inc.
- Schweitzer Industries Corporation
- TEC Systems

Letters were sent to the vendors requesting information on their particular catalytic incinerator systems. Replies from six of the nine vendors were received. Their replies are summarized below.

---

*Subsequent to the writing of this report, DuPont sold its Torvex^R catalytic incinerator business to Englehard.

Air Resources, Inc. markets fluidized bed catalytic incinerators. The catalyst, developed by Harshaw Chemical Company, is a nonprecious metal disposable pellet. The fluidized design overcomes masking problems. The moving catalyst bed allows masked catalyst to move to higher temperature zones in the bed, thereby oxidizing the masking material. About fifteen of these fluidized bed units are used for VOC control. Of the fifteen, a signi- ficant portion are used for control of chlorinated hydrocarbons.[25]

CE Air Preheater markets both thermal and catalytic incinerator systems in sizes ranging from 2,000 scfm to over 100,000 scfm. Their catalyst, supplied by Johnson Matthey[27], is a platinum deposited, stainless steel honeycomb. CE Air Preheater has installed approximately 10 catalytic inciner- ator systems since the early 1970's.[26]

E.I. DuPont de Nemours & Company, Inc. markets catalytic incinerator systems in sizes ranging from less than 1,000 scfm to 70,000 scfm. Their catalysts consist of precious metal deposited on ceramic honeycombs. Since 1976, DuPont has supplied about 100 complete catalytic incinerator systems and retrofitted about 200 thermal incinerators with catalyst.[28]

Engelhard, a leading supplier of automotive catalysts, markets catalytic incinerator systems ranging in size from less than 100 scfm to over 100,000 scfm. Their catalyst is a precious metal deposited on a honeycomb support.[29]

Johnson Matthey, Inc. has provided both catalysts and complete catalytic incineration systems for various applications throughout the world. Currently in the United States, Johnson Matthey, Inc. supplies catalyst to CE Air Preheater.[27]

Met-Pro Corporation, formerly Oxy-Catalyst, markets both thermal and catalytic incineration systems and carbon adsorption systems. They market a wide range of catalysts including both precious and nonprecious metals deposited on a monolithic or pellet support. They have installed over 200 catalytic incinerator systems throughout the United States and some foreign countries.[30]

TEC Systems manufacturers and markets catalytic incinerator system that utilize a grandular catalyst bed. These incinerators have been effectively used to control VOC emissions from a number of industrial processes.

3.2 STATUS OF DEVELOPING TECHNOLOGY

For the purpose of this report, a developing technology is considered to be a process, operation or equipment piece not yet in full scale commercial use.

Catalytic incineration is a reasonably mature technology that has been in use since the 1940's. Although improvements in catalysts have made the technology less costly and more energy efficient, no major developments in the technology have occurred since the early 1970's.

Fluidized bed catalytic incineration was a new development in the early 1970's. Developed jointly by Air Resources, Inc. and Harshaw Chemical Company, this technology has been in commercial use since 1975 and is being marketed on a turnkey basis.[31]

Catalyst and catalytic incinerator vendors are continually researching improvements in catalyst characteristics and formulations. One catalytic incinerator vendor is also concentrating development effort into energy efficiency and heat recovery.[32]

Industrial users of catalytic incinerator systems are currently attempting to optimize waste gas flow rates and VOC concentrations. Most drying ovens maintain LEL levels at 25 percent or less to insure the process excursions remain comfortably under the LEL. Use of a VOC monitor on the drying oven exhaust allows users to operate ovens at higher LEL levels (greater than 25%) with relative safety. The increase in VOC concentration in the ovens increases waste gas VOC concentrations thereby making incinerators more energy efficient.

3.3 APPLICATIONS OF CATALYTIC INCINERATORS

This section presents and describes the processes to which catalytic incinerators are being applied to control VOC emissions. It begins with an overview of the major stationary sources of VOC emissions. Following this is a subsection listing the industrial processes using catalytic incineration. This list is discussed in more detail in the subsequent subsections which group applications of catalytic incinerators into three categories:

- Solvent evaporation processes
- Organic chemical manufacturing processes
- Miscellaneous processes

3.3.1 Overview of VOC Emissions from Stationary Sources

Volatile organic compounds are emitted from a variety of stationary sources. Total nationwide VOC emissions for 1975 were estimated to be 31 million tons, of which 19 million tons were from stationary sources.[33] Table 3-9 summarizes these stationary sources and their percentage contribution to the total national VOC emissions from all stationary sources based on 1972 data.

TABLE 3-9. STATIONARY SOURCES AND THEIR RELATIVE CONTRIBUTION TO NATIONWIDE STATIONARY SOURCE VOC EMISSIONS[34]

Stationary Source Category	National Emissions (%)
Solvent evaporation	48.0
Refuse disposal	11.6
Agricultural burning	11.2
Refineries	11.0
Gasoline marketing	6.1
Carbon black manufacture	4.8
Wood burning	2.6
Charcoal manufacture	1.4
Metallurgical coke manufacture	1.3
Fuel oil burning	1.2
Organic chemical manufacture	0.6

Catalytic incineration has been applied to waste streams in several of these stationary source categories. Most applications are in the categories of solvent evaporation and organic chemical manufacture, with limited use in several other categories.

3.3.2 Current Applications of Catalytic Incinerators

It has been estimated that more than 500 catalytic incinerators are in use today. Only 149 plants using catalytic incinerators have been identified in this study. Table 3-10 summarizes the industries identified in this study to which catalytic incinerators are being applied. This table is derived from literature and from vendors. These information sources either cite specifically the actual plants in which catalytic incinerators are being used or state that catalytic incinerators are applicable to a particular industry. Information identifying actual plants can be found in Chapter 7.

The list is divided into three categories: solvent evaporation processes, organic chemical manufacturing, and miscellaneous sources. Solvent evaporation processes are by far the major stationary source of VOC emissions, and catalytic incineration is widely used in many of the industries within this category. Organic chemicals manufacturing contributes substantially less VOC emissions. However, there is some information describing the application of catalytic incinerators to the manufacturing of some of these chemicals. The miscellaneous category includes those sources for which there is little readily available information concerning the applicability of catalytic incinerators. The following subsections discuss these categories in more detail.

TABLE 3-10. CURRENT APPLICATIONS OF CATALYTIC INCINERATORS[36]

1. Solvent Evaporation Processes

   Can coating
   Metal coil coating
   Auto and light-duty truck coating
   Metal furniture coating
   Wire enameling
   Miscellaneous metal parts coating
   Fabric coating
   Paper coating
   Graphic arts

2. Organic Chemical Manufacturing

   Ethylene oxide
   Acrylonitrile
   Formaldehyde
   Ethylene dicholride and vinyl chloride monomer
   Cumene
   Caprolactam
   Maleic Anhydride

3. Miscellaneous Processes

   Plywood manufacturing
   Gasoline marketing
   Foundry core baking
   Tire manufacturing
   Varnish cooking
   Textile processing
   Kraft paper manufacturing
   Carbon baking
   Filter paper processing
   Catalyst regeneration
   Asphalt blowing

Refer to Appendix B for complete list of known plants.

3.3.3 Solvent Evaporation Processes

In 1972, it was estimated that 48 percent of all emissions from stationary sources were from the evaporation of organic solvents.[6] Organic solvents are used to reduce the viscosity of paints, enamels, lacquers, and inks to facilitate their application in surface coating and printing operations. These solvents vary substantially among coating operations and often consist of a mixture of organic compounds. In many instances the formula is proprietary information. Table 3-11 contains a listing of the most common chemicals used as solvents.

TABLE 3-11. SURFACE COATING SOLVENTS[37,38]

Toluene	Ethanol
Xylene	Butanol
Hexane	Gycols
Isoctane	Gycol ether esters
Mineral spirits	Gycol esters
Naptha	Acetone
Heptane	Methyl ethyl ketone
Propanol	Methyl iso-butyl ketone
Isopropanol	Isopropyl acetate
Methanol	N-propyl acetate
Cyclohexanone	Ethyl acetate

The average solvent consumption per gallon of solvent coatings has decreased due to the effects of air quality standards which regulate organic solvent emissions and due to the trend toward using more efficient, high solids industrial coatings. The growth of solventless application methods such as powder coatings and coatings with reduced organic solvent concentration such as waterborne solvents have also decreased solvent consumption. Table 3-12 demonstrates the trend toward lower usage of organic solvents.

TABLE 3-12. ORGANIC SOLVENT VS. WATER USAGE IN SURFACE COATINGS

Type	1966	1969	1973	1975
Organic solvents	77.5%	76.1%	74.8%	72.4%
Water	22.5%	23.9%	25.2%	27.6%

In surface coating operations, the majority of emissions are released during application and during curing. Application techniques include spray booths, dip tanks, flow coaters, roll coaters, and printing presses. Curing is usually accomplished in a bake or cure oven.

Catalytic incinerators are typically more cost effective when applied to cure ovens. In fact, most catalytic incinerators that have been installed in surface coating operations have treated cure oven exhaust rather than applicator station exhaust. In contrast to applicator station exhaust, oven exhaust normally has a much lower flow rate, requires less preheat, is relatively free of particulate matter, and does not require a system to capture the VOC vapors. In addition, waste gases from cure ovens generally have higher concentrations of VOC's and are at higher temperature than waste gases from applicator stations. This means that the fuel cost for treating oven exhaust is significantly less than the fuel cost for treating applicator station exhaust.

In some coating operations, problems may be encountered with the use of catalytic incineration systems. The major problem is catalyst deactivation. Catalysts are deactivated when exposed to trace amounts of metals such as lead, antimony, zinc, and copper. Some of these metals are present in the pigment component of some coatings. In addition, the catalyst may be masked by high molecular weight organics, alumina, and silica dust and suppressed by halogens and sulfur, each of which is present in some coatings. Because of this problem of deactivation, catalytic incineration has been most suitable for captive coaters who coat only a few different products with a limited number of coating formulations. These coaters can insure that no chemical poisons are present in the coatings. Commercial coaters, however, who must often use a wide variety of coatings specified by their customers, do not have this control.

Table 3-13 summarizes oven exhaust parameters for various coating operations. Process descriptions, solvent usage, the applicability and current use of catalytic incinerators, and the potential for heat recovery using catalytic incinerators are discussed in the following subsections.

Can Coating

Approximately 50 percent of all cans produced are used as beverage containers. The data presented in Table 3-13 for can manufacture represent emissions from the beverage can coating industry.

Beverage cans are made in two-piece and three-piece styles. For each, an exterior base coat is applied followed by a lithography operation. An overvarnish is often applied to protect the lithography and improve the appearance of the can. Finally, a protective coating is applied to the inside to protect the purity and flavor of the can contents.

TABLE 3-13. SUMMARY OF OVEN EXHAUST STREAM PARAMETERS FOR SOLVENT EVAPORATION[26]

Industry	Entering Temperature (°F)	Gas Flow Rate (scfm)	Gas Composition (ppm)	(%LEL)	Fuel Value (Btu/scf)
Can Manufacture (Beverage)	225-425	2,000-14,000	1200-3600	5-15	2.5-7.5
Metal Coil	600	5,000-33,000	3000	21	14.5
Auto and Light-Duty Truck	200-450	1,600-25,000	1200-2400	5-10	2.5-5.0
Metal Furniture	200-300	4,000-35,000	1500	6	3.0
Wire Enameling	500-850	1,000-15,000	2400-6000	10-25	5.0-13
Miscellaneous Metal Parts	275-450	2,000-78,000	600-6000	3-25	1.5-12.5
Textile Coating	300	3,000-30,000	700-7000	5-50	3.0-30
Paper Coating: Pressure Sensitive Tape and Label	200	200-30,000	3500-7000	25-50	15-30
Graphic Arts:[a] Rotogravure and Flexography	105	50,000-130,000	1000-2000	15-20	7.5-10
Vinyl Coating	300	2,000-80,000	700-2200	5-25	2.5-13

[a]From both ovens and applicator stations.

In this industry, solvent is evaporated from the application process and from the cure ovens. The relative contribution from each varies among the coating operations. For example, cure oven emissions represent 22 percent of the VOC emissions for a plant coating can bodies for two-piece cans; however, cure oven emissions comprise 90 percent of total emissions from a plant coating steel bodystock for three-piece cans. Typically, oven exhaust is 5-15 percent LEL while applicator station exhaust is 0.5-2.5 percent LEL. Currently, catalytic incineration is used only for oven exhaust in the beverage can coating industry.[42]

A large number of diverse solvents are used in both exterior and interior coatings. The trend is to use waterborne coatings instead of solvent-based coatings. In 1978, it was estimated that waterborne coatings comprised between 15 and 25 percent of total solvents used in beverage can coating.[43] In fact, the use of waterborne coatings is currently the most common control technique.[43]

Catalytic incineration is widely used in can coating operations. It has been used by the major can manufacturers at an increasing rate over the last seven years.[44] The incinerator's exhaust stream is used to preheat the incoming effluent and/or to preheat the cure oven air either directly or via heat exchange. A few plants use the heat to produce hot water for line-washers, hot air for dry-off ovens, or they use it for space heating.[44]

Coil Coating

Coil coating is the prefinishing of sheet metal with protective and decorative coatings that are applied by roll coating on one or both sides of a fast-moving metal strip. The metal strip unwinds from a coil and travels through a series of roll coaters before passing through a series of cure ovens. The ovens maintain higher temperatures (500- 700°F) than those in most other coating operations (200-400°F) because of the metal strip's short residence time. After baking, the strip is rewound into a coil. Eventually, the metal is used in many products including cans, appliances, industrial and residential siding, shelving, cars, gutters, and downspouts.[45]

Approximately 90 percent of VOC emissions in the coil coating industry come from the curing ovens.[46] These exhaust fumes were catalytically incinerated 20 years ago. However, the advent of siliconized coatings for weather durability decreased the application of catalytic incinerators because of the severe masking problems associated with them. New catalyst developments and increasing natural gas costs have renewed the interest in catalytic incinerators.

Approximately 85 percent of the coatings used in the coil coating industry are solvent based. The rest are waterborne solvents which contain 2-15 volume percent organic solvent.[47] The waterborne coatings generally produce a lower mass of VOC emissions per unit of coating solids applied, but they have not as yet been developed for all end-product uses.

Unlike many other coating operations, the oven exhaust in coil coating can typically be incinerated catalytically without preheat. Further energy savings can be realized for a small capital investment by recycling a portion of the incinerator's hot exit gas back to the ovens. At least four facilities have incorporated this into their catalytic incineration systems.[48]

Auto and Light-Duty Truck Coating

Auto and light-duty truck assembly plants and their related subassembly parts plants apply large quantities of coating in spray booths. From 70-90 percent of total plant VOC emissions are exhausted from these booths.[47] The large flow rates (200,000-400,000 scfm)[49] from these stations preclude the application of catalytic incineration for VOC control because of the enormous catalyst volume required. Since most of the solvent vapor is dissipated prior to the bake ovens, the ovens also emit relatively low concentrations of VOC. The two major automobile and light-duty truck manufacturers report a VOC concentration of 5 percent LEL for oven exhaust. More catalyst and fuel are required per unit flow rate in these coating operations than in other coating operations. As a result, catalytic incineration is not routinely used in the automotive industry at this time.[50]

In the past, catalyst beds were used widely by the auto industry as an integral part of the cure oven, not as a supplemental VOC control device. These beds were used primarily to prevent fires and limit condensation of organics on the oven walls.

However, the need for any type of incineration, has almost been eliminated by the advent of waterborne paints whose emissions easily meet typical hydrocarbon regulations.[48] In fact, waterborne coatings are currently the most common VOC control technique in use in the vehicle assembly industry.

Metal Furniture Coating

The metal furniture coating industry uses primarily solvent-based coatings which are applied primarily by spray and dip coating processes. These coatings typically contain 65 percent solvent.[51] After the furniture is pretreated to remove oil and grease and to improve the adhesion characteristics of the surface, it passes through an applicator station where it is coated. Then it passes through a flash-off area before curing in a bake oven. As in other coating operations, catalytic incineration is most cost effective when applied to oven exhaust.

Electrostatic spraying is the most common application technique. Approximately 50 to 95 percent of the solids adhere to the surface of the furniture.[52] The transfer efficiency depends on the application equipment and the configuration of the item being painted. Seventy percent of the solvent in the coating evaporates from the spray booths and flash-off areas, with only 30 percent remaining to be evaporated by the ovens. Because of the small solvent concentration in the oven exhaust, catalytic incineration is typically not used in those operations employing spray booths.

Dip coating, on the other hand, has a much higher transfer efficiency. This means that more of the solvent remains in the coating in dipping than in spraying. Therefore, 50 to 70 percent more solvent is evaporated from the baking ovens when dipping is employed than when spraying is employed. Thus, catalytic incineration is more applicable to those processes using dipping.

Waterborne coatings are becoming more popular in the metal furniture coating industry. The organic solvent content in these range from 10-50 weight percent.

Wire Enameling

Wire enameling involves the application of an electrically insulating varnish or enamel to an aluminum or copper wire for use in electrical machinery. As in many other coating operations, multiple coats are applied, each time passing through a cure oven for curing.

The organic solvent content of wire coatings ranges from 67 to 85 percent by weight. Cresylic acid, various cresols, xylene, and mixtures of C_8 - C_{12} aromatics are widely used.[53]

Solvent vapors are predominantly emitted from the cure ovens. Their high operating temperatures (greater than 650°F) and their moderate to high VOC concentrations make catalytic incineration the most viable control method.

Catalytic incineration can be applied internally and externally. Internal catalytic incinerators were first introduced in the late 1950's. Today, all major manufacturers of wire coating ovens incorporate metal catalysts into their designs.[54] Reduction of oven fuel requirements is the primary reason for using internal catalysts whereas pollution control is only a secondary consideration. Older ovens without an internal catalyst are usually equipped with an external catalyst. In these cases, the catalyst system is added primarily for pollution control since the heat is not as easily recovered.

Miscellaneous Metal Parts Coating

Miscellaneous metal parts coating is a broad category summarizing those metal coating operations that have not previously been discussed. It includes large and small farm machinery, commercial machinery, industrial machinery, small appliances, and fabricated metal parts. There are far more dissimilarities than similarities among the industries represented by this category and even among the various plants within each industry. However, an attempt has been made to generalize the processes and their pollution characteristics.

Spraying is the most common method of applying coatings to miscellaneous metal parts for single coat operations, but flow coating and dipping are also used. For two coat operations, the first coat is more likely to be applied by flow or dip coatings while the topcoats are almost always sprayed.

As in many other coating operations, the use of waterborne coating is a viable VOC control alternative. It can reduce VOC emissions 60-90 percent.[55]

Fabric Coating

Textile or fabric coating involves the coating of a textile substrate to impart properties not originally present in the fabric such as strength, water repellancy, and stability. As with other coating facilities, a fabric coating line consists of an application area and a drying oven. Between 40 and 70 percent[56] of VOC emitted from the plant are emitted from the coating line of which 65-75 percent are emitted by the drying ovens.[57] The data in Table 3-13 are for the oven exhaust stream.

Although the range of oven exhaust flow rates varies widely, it is typically 10,000-20,000 scfm. Old plants with semi-enclosed ovens have solvent concentrations of 5-12 percent LEL in the oven exhaust whereas newer plants have concentrations above 25 percent LEL.[58] At least 3 plants in the U. S. are operated at 40-50 percent of the LEL.

VOC emissions can be reduced from 80-100 percent[59] by using coatings with low organic solvent concentrations. Although many facilities have converted to these coatings, the coatings are not a universal control measure. Solvent-based coatings still have distinct advantages in many applications.

Paper Coating

Paper is coated for a variety of decorative and functional purposes. The majority of VOC emissions from paper coating come from the coating of pressure sensitive tapes and labels.[60] For this reason, the data from these operations are used to represent the paper coating industry. In these operations, a backing material such as paper, cloth, or cellophane is coated several times with an adhesive coating to create a tape or label that sticks on contact.

More than 85 percent of the total production of pressure sensitive materials are produced using solvent-based coatings.[61] Eighty to 95 percent of all VOC emissions from these coating operations are contained in the drying oven exhaust.[62] The exhaust from older lines are controlled at 25 percent LEL while newer lines have increased this to 40-50 percent LEL.[63] The exhaust flow rates range from 200 scfm for a small silicone releasing coating to approximately 30,000 scfm for large adhesive coating lines.[64]

Although catalytic incinerators have been applied to these oven exhaust streams, there is little data quantifying its application.

Graphic Arts

The graphic arts industry consists of printing operations which fall into four principal categories: letterpress, offset lithography, rotogravure, and flexography. VOC emissions from rotogravure and flexography together account for 54 percent of VOC emissions from the total printing industry.[65] Data from these operations are used to represent the printing industry at large and are compiled in Table 3-13.

A vast majority of the printing units use organic solvent-based inks. Rotogravure inks contain 50-85 volume percent organic solvent, while flexographic inks typically contain 75 volume percent organic solvent.[65]

Catalytic incineration is being used in the flexography and packaging rotogravure printing industries. However, it is seldom being applied if at all in rotogravure publication operations. In these operations, carbon adsorption is the most widely used VOC control technique.

Substitution of waterborne inks for solvent-based inks can achieve emission reductions comparable to those achieved by catalytic incineration. However, solvent-based inks are still required in many applications.

3.3.4 Organic Chemical Manufacturing

This study has found that within the organic chemical manufacturing industry incineration is applied predominantly to processes using air oxidation. The VOC emissions from air oxidation chemical plants vary among the chemicals produced and many times between two different processes that manufacture the same chemical.

Nearly all oxidation processes (either air or oxygen) are exothermic. The heat from their reactions is generally used to produce steam. The products are partially condensed and the remaining volatiles are vented to an absorber. The absorber uses some liquid, usually water, to transfer the desired product from the vapor to the liquid phase. And the waste gas from the absorber is typically the major source of VOC emission from the process. It is typically vented at 100°F.[66] An abridged flow diagram of a typical air-oxidation process showing the location of the absorber is shown in Figure 3-4. Table 3-14 lists waste gas characteristics from various oxidation processes.

Ethylene Oxide

In the United States, two major processes are used in the manufacture of ethylene oxide. Sixty-three percent is manufactured at six locations by air oxidation of ethylene, and 37 percent is manufactured at eight locations by oxygen oxidation of ethylene.[68] In the air oxidation process, ethylene oxide is separated from the other reaction products in an absorber. The unabsorbed gas is vented and contains predominantly nitrogen, oxygen, ethylene, ethane, and carbon dioxide. This stream accounts for 90 percent of uncontrolled VOC emissions from the air oxidation process.[69] Currently only catalytic oxidation is being used for control of VOC emissions from this source.[70] Its primary function is not in pollution abatement; rather, it is an integral part of a system that recovers energy from the process off-gases by using turbines for compression of the systems process air feed.[70]

There is a major restriction that must be considered with catalytic incineration. Ethane is more difficult to burn than ethylene, because it has a higher ignition temperature. This difference in burning characteristics requires that the ethylene-to-ethane ratio in the feed be controlled to ensure a high enough ethane ignition temperature and yet ensure that the temperature limitation of the catalyst is not exceeded.

In the oxygen-oxidation process, there are two major sources of VOC emissions: the CO_2 purge vent and the argon purge vent. Control of the CO_2 purge stream by catalytic incinerators is considered impractical,[71] and the argon purge system can sustain combustion and can thus be used as a fuel gas.

Acrylonitrile

All six acrylonitrile plants in the U. S. use the SOHIO process. Air, ammonia, and propylene react in a fluidized bed reactor and the products are cooled in a quencher. Volatiles from the quencher go into an absorber where the acrylonitrile product and by-products (hydrogen cyanide and acetonitrile) are generally vented directly to the atmosphere. However, in one plant a thermal incinerator is used and in another a catalytic incinerator is used. This absorber vent stream is one of two where hydrocarbon emissions present a serious pollution problem. The other, evaporation from a deep well pond, is not suitable for catalytic incineration. Catalytic incineration has only been applied to the absorber vent stream.[72]

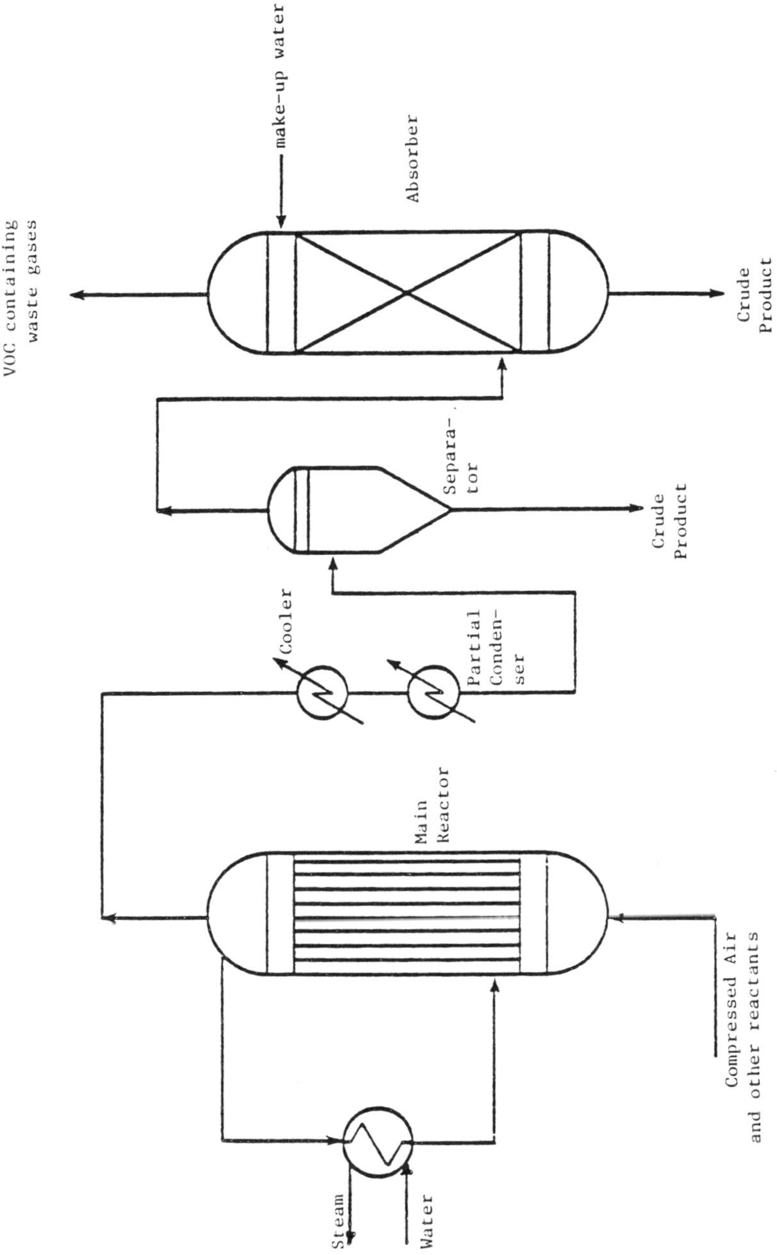

Figure 3-4. Flow diagram of the initial stage of an air-oxidation process.

TABLE 3-14. PARAMETERS OF MAJOR VOC EMITTING WASTE GAS STREAMS IN PETROCHEMICAL MANUFACTURING[26]

Petrochemical	Process	Gas Flow (scfm)[a]	Gas Composition VOC (ppm)	Fuel Value[b] (Btu/scf)
Ethylene Oxide				
Air Oxidation	Absorber	30,000	9,100	16
Oxygen Oxidation	Argon Purge	100	280,000	550
Acrylonitrile	Absorber	90,000	6,800	16
Formaldehyde				
Silver Process	Absorber	1,700	4,000	82
Metal Oxide Process	Absorber	2,700	2,100	6.9
Ethylene Dichloride Vinyl Chloride Monomer	Oxychlori-nation Reactor	8,700	11,000	23
Maleic Anhydride				
from Benzene	Absorber	28,000	1,400	11
from NButane	Absorber	21,000	2,000	19

[a]Standard temperature is 60°F.

[b]Includes other combustible gases in addition to VOC's such as methane, hydrogen, and carbon monoxide.

The major hydrocarbons emitted by the absorber vent are propane, proplyene and acetonitrile with acetonitrile comprising the majority of the total absorber vent VOC.

Formaldehyde

Two major processes are used in the United States to manufacture formaldehyde from methanol: 75 percent is produced by the silver process which is used at 35 locations and 25 percent is produced by the metal-oxide process which is used at 20 locations. The silver process involves the dehydrogenation of methanol followed by the oxidation of a portion of the hydrogen evolved to form water. The metal-oxide process involves oxidation of methanol. The major difference between the two processes is the amount of air mixed with the methanol before conversion. The silver process operates above the upper explosive limit, whereas the metal-oxide catalyst process operates below the lower explosive limit.

As in the ethylene oxide process, the product gases are separated in absorbers. In the silver process, the absorber vent gas comprises 62 percent of the total VOC emitted from the process and the vent gas from the formaldehyde/methanol distillation column comprises 21 percent.[73] The major VOC emitted in the absorber vent gas are formaldehyde, methanol, methyl formate, and methylal. The stream also contains 20.25 percent hydrogen, 0.36 percent carbon monoxide and 0.45 percent methane which are all combustible and in these concentrations raise the heat value of the stream substantially.[74] The metal-oxide process has a higher yield than the silver process and thus does not require distillation. The absorber vent gas comprises 80 percent[75] of the VOC emitted from this process. The major VOC emitted are formaldehyde, methanol, and dimethyl ether. Carbon monoxide is also emitted at 12,500 ppm.[76]

The absorber vent stream from the silver process is rich in hydrogen gas and contains clean-burning hydrocarbons. The heating value of the gas is high enough to self-sustain combustion. Because of this, catalytic incineration is not applied.

The vent gas from the metal oxide process is largely inert and cannot sustain combustion. Its fuel value is about 7 Btu/scf[67] as opposed to the silver process's 80 Btu/scf.[67] Therefore, catalytic incineration is more applicable to the metal-oxide process than to the silver process.

Ethylene Dichloride and Vinyl Chloride Monomer

Most of the ethylene dichloride produced in the United States is used in the production of vinyl chloride monomer (VCM). Three principal stages are involved in producing VCM. First, ethylene dichloride is made by the direct chlorination of ethylene. This compound is then cracked to yield VCM plus hydrogen chloride. Finally, the HCl and oxygen are reacted with ethylene via an oxychlorination reaction to produce additional ethylene dichloride. Air typically provides the oxygen source in this oxychlorination step. These steps constitute the "balanced" process, so-called because all of the HCl is recycled.[77] Approximately 95 percent of all ethylene dichloride produced domestically is by this balanced process.

In the oxychlorination reaction, ethylene dichloride is recovered in an absorber. The waste gases are typically vented to the atmosphere. In the direct-chlorination step of the balanced process, the products are immediately sent to a cooler. There, any inert gas fed with the reactants is released to the atmosphere. Further down line, more gases are vented from a series of distillation columns.

Catalytic incineration has been used to control emissions from an oxychlorination absorber vent. It has removed carbon monoxide and ethylene with greater than 99.7 percent reduction; however, it has removed less than 75 percent of the ethylene dichloride and less than 60 percent of the vinyl chloride monomer.[78] Because of the large percentage of nitrogen and carbon dioxide in the vent gases, supplemental fuel typically must be added for proper combustion. No information was available concerning applicability of catalytic incinerators to the direct chlorination and column vent streams. It is known that thermal incinerators are being used.

Maleic Anhydride

Catalytic incineration has been applied commercially since 1978 to the manufacture of maleic anhydride.[79,80] Waste gas characteristics are presented in Table 3-14.

Currently, 85 percent of maleic anhydride is produced by vapor phase air oxidation of benzene and 15 percent by butane oxidation.[81] As of 1978, there were eight domestic facilities producing maleic anhydride in ten plants. A distinct trend has been observed towards replacing benzene with butane as the feedstock because of its availability and lower cost. Overall VOC emissions for the butane process are about twice that of the benzene oxidation process. However, all process data concerning the butane process are currently proprietary and unavailable.[82]

In the benzene oxidation process, crude maleic anhydride is received in an absorber where it is contacted with water or aqueous maleic acid. The absorber's waste gas stream accounts for 97 percent of the benzene and total VOC emitted from the plant.[82] The emissions are influenced by the excess air fed to the reactor to maintain the benzene concentration below the explosive limit.

One company is using a catalytic incinerator to control VOC emissions from this absorber. The maximum practical achievable VOC removal efficiency is reported to be less than 95 percent.[83]

Other Organic Chemicals

Caprolactam and cumene are two other organic chemicals identified as using catalytic incineration. In caprolactam manufacture, the only source to which catalytic incineration is applied is the dehydrogenation reactor.[66] The vent stream from this reactor can contain significant VOC and hydrogen. However, data on the composition and flow are not available. Normally, this stream is burned as fuel.[85]

In cumene manufacture, the vast majority of VOC is fugitive or comes from storage and handling. In the phosphorous acid catalyst process, the only stream catalytic incineration could be applied to is the cumene distillation vent stream. It contains 35.9 percent VOC and represents 6.2 percent of all VOC emitted from the process. It is normally flared. In the aluminum chloride catalyst process, there are five streams to which catalytic incineration would apply. Mixed together they contain 32.5 weight percent organics and represent 10.8 weight percent of all VOC emissions from the plant.[85] They, too, are normally flared.[87]

3.3.5 Miscellaneous Sources

There are other sources that our investigation reveals are using catalytic incinerators. They have been used to control emissions in varnish cooking, catalyst regeneration, foundry core ovens, textile processing, rubber manufacturing, carbon baking, plywood manufacturing, gasoline marketing, Kraft paper manufacturing, filter paper processing, and tire manufacturing. Some are touched upon briefly in the following descriptions.

Varnish Cookers

Varnish cooking processes modify natural or synthetic oils or resins which are the film-forming vehicles in inks or coatings (i.e., varnish, paint, enamel, lacquer). Many new coatings do not use these film-forming products, however, as a result, the products of varnish cooking constitute a much lower percentage of all coating materials than they did a decade ago. Nevertheless, they are still extensively produced for use in the manufacture of surface coatings.

Some contaminants released in a varnish cook are fatty acids, glycerine, acrolein, phenols, aldehydes, ketones, terpenes, hydrogen sulfide, and particulates. Cooker emissions vary in composition depending on the ingredients in the cook. Typically, VOC concentrations range from 200 ppm - 2000 ppm, and particulate concentrations range from 0.30 - 0.64 gr/scf. Also, the volumetric flow rate and temperature of the waste stream are 400-1,000 scfm and 100-120°F, respectively.[88]

Catalytic incinerators are known to have been installed to control emissions from varnish cookers. Particulate loading may be a problem if the exhaust is not precleaned before incineration.

Catalyst Regeneration

Catalysts for the catalytic cracking and reforming processes in refineries are regenerated to restore activity by burning off the coke and other deposits from the catalyst surface.

Air contaminants are released to the atmosphere during regeneration, especially when combustion is involved. Types of contaminants emitted are catalyst fines, oil mists, hydrocarbons, ammonia, sulfur oxides, chlorides, cyanides, nitrogen oxides, carbon monoxide, and aerosols.[66] The pollutants released by any one type of regenerator depend on the compositions of the catalyst and reactants, and the operating conditions. Care must be taken in such operations to remove the catalyst fines and sulfur compounds which could deactivate the catalyst in a catalytic incinerator.

Foundry Core Ovens

In foundries, core ovens are used to bake the cores used in sand molds. Generally, emissions from core ovens are minor when compared to other metallurgical processes. VOC are emitted as the binders and resins in the molds are cured. The contaminants discharged include organic acids, aldehydes, hydrocarbons, and particulates. VOC concentrations typically range from 200-600 ppm. The exhaust gas volumetric flow rate and temperature are 100-250 scfm and 400°F, respectively.[44]

Incinerators used to control core oven emissions are predominantly thermal incinerators. In some instances, particularly in large core ovens, catalytic incinerators have been used. When they are used, care must be taken to keep the catalyst in good condition; otherwise, partial oxidation can result in the discharge of combustion contaminants more objectionable than the oven effluent. A recent trend is toward inorganic binders and resins in lieu of organic binders and resins. For this reason, it is unlikely that incinerators will be installed on new foundry core ovens.

Textile Processing

No current uses of catalytic incinerators in the textile industry were identified. However, several emission sources could possibly be controlled using catalytic incineration and efforts are being made to test catalytic incinerators on these sources. A most promising application is the waste stream from a Suessen heat setting device. There are approximately 250 such devices in use. Typically the waste stream from this device is 250-300°F, has a flow rate of 1200 scfm, contains 0.02 weight percent oil, and has a heating value of 0.83 Btu/scf. Despite this relatively low fuel value, a heat recovery system under development makes catalytic incineration economical.[89]

Kraft Pulping Processes (Paper Manufacture)

Catalytic incineration within the Kraft pulping process has been limited to pilot-size trial projects.[79] The stream on which it has been demonstrated is the vent stream from the sparging of black liquor. A 95 percent reduction in concentration of odorous organic sulfides and of wood-source hydrocarbons has been attained. No information was available concerning the effect the sulfur compounds have on the catalyst.

Plywood Veneer Drying

A plywood mill in Oregon has applied catalytic incinerators to control the fumes generated in the drying of plywood veneers.[90] There were initial problems with tiny wood splinters blocking the catalyst inlet; however, raising the temperature of the inlet stream to 400-600°F pyrolyzes the splinters and allows the unit to function relatively maintenance free.

Filter Paper Processing

In making fuel oil and air filters for vehicles such as motorcycles and diesel locomotives, heat processing of the filter paper is necessary to cure the resin with which the paper was impregnated. The oven exhausts contain between 10-20 percent LEL VOC, with alcohols and resin monomers being the primary pollutants.[91] The resins sometimes contain organic phosphate fire retardants which deactivate the catalyst. Mild acid leaching has restored activity in at least one case.

Catalysts are being installed internally in the majority of new filter paper cure ovens as an integral part of the oven recirculation/burner system.[92] They keep the oven interior free of condensed resins and eliminate the opacity and odor in the exhaust. They also reduce oven fuel requirements.

Gasoline Marketing

A system marketed under the trademark Vapox eliminates fugitive vapors from gasoline filling stations and from bulk loading stations.[92] The incinerator contains a platinum catalyst on a honeycomb bed.

REFERENCES

1. Kirk-Othmers Encyclopedia of Chemical Technology. Third Edition: New York, Wiley Interscience, 1980, Vol. 9. pp. 511-541.

2. Kenson, R. E., and W. D. Larlee. Catalytic Incineration - the Fuel Saver. (Presented at the North California Section of the Metal Decorators Association. 1978) 23 p.

3. Rolke, R. W., et al. (Shell Development Company.) Afterburner Systems Study. (Prepared for U. S. Environmental Protection Agency.) Research Triangle Park, N.C. Publication No. EPA-R2-72-062. August 1972. pp. 55-83.

4. Meinhold, T. F. Fume Incinerators for Air Pollution Control. Plant Engineering. 34:108-115. November 13, 1980.

5. Kroehling, J. H. (DuPont). Catalytic Oxidation; An Energy Saving Approach to Fume and Odor Control. (Presented at the Specialty Conference on State-of-the-Art of Odor Control Technology II, Air Pollution Control Association. Pittsburgh. March 1977). 11 p.

6. Kent, R. W. A Guide to the Control of Volatile Organic Emissions, Harleysville, Pennsylvania, Met-Pro Corporation, n.d. 7 p.

7. Martin, Norman A. (Englehard Minerals & Chemicals Corporation.) Catalytic Incineration of Low Concentration Organic Vapors. (Prepared for U. S. Environmental Protection Agency.) Research Triangle Park, N.C. Publication No. EPA-600/52-81-017. October 1981. p. 4.

8. Searles, R. A. Catalysts Cut Costs by Reducing Temperature for Incinerating Plant Odors. Process Engineering. 54:100-102. February 1973.

9. National Tire Protection Association. Fire Hazard Properties of Flammable Liquids, Gases, Volatile Solids, 1977. NFPA-325M, Boston, MA. 1977. p. 325M-15.

10. Letter from Kashmer, B. A., Johnson Matthey, Inc., to Krohn, N. E., Radian Corporation. October 10, 1981. 2 pp. Information on Catalysts.

11. Telecon. Madden, G. I., DuPont with Krohn, N. E., Radian Corporation. January 7, 1982. Conversation about catalyst tolerances.

12. Reference 3, pp. 89-96.

13. Vincent, E. J., J. C. Berry, and D. E. Young (EPA:OAQPS). Are Afterburners Obsolete? (Presented at meeting of National Association of Corrosion Engineers.) Atlanta, GA. January 19, 1978. 16 p.

14. Letter from Pennington, R. L., Reeco, to Krohn, N. E., Radian Corporation. October 7, 1981. 3 pp. Information about the REECO System.

15. Juhola, A. J. (MSA Research Corporation.) Package Sorption Device System Study. (Prepared for U. S. Environmental Protection Agency.) Evans City, PA. Publication No. EPA-R2-73-202. April 1973. Chapters 5 and 7.

16. Memo from Krohn, N. and M. Jennings, Radian Corporation, to file. July 30, 1981. Notes of meetings with N. Martin and Other Englehard representatives and EEA and EPA representatives.

17. Letter and attachments from Dey, H. F., South Coast Air Quality Management District, to Krohn, N. E., Radian Corporation. November 16, 1981. 71 pp. Transmittal of five permits for catalytic afterburner installations and seven source test reports.

18. Letter and attachments from O'Malley, C. A., South Coast Air Quality Management District, to Berry, R., Radian Corporation. November 19, 1981. 20 pp. Information on source tests conducted on catalytic incinerators.

19. Trip Report. Key, J. A., Hydroscience, Inc., to file. September 7, 1977. 43 pp. Report of visit to E. I. duPont de Nemours & Company in Beaumont, Texas to observe emission control options for the synthetic organic chemicals manufacturing industry.

20. Letter and attachments from Jennings, R. L., State of Illinois Environmental Protection Agency, to Krohn, N. E., Radian Corporation. December 22, 1981. 10 pp. Information on catalytic incinerators in Illinois.

21. Letter and attachments from Brown, D. L., State of Ohio Environmental Protection Agency, to Krohn, N. E., Radian Corporation. October 13, 1981. 39 pp. Information on the use of catalytic incineration to control emission of volatile organic compounds in Ohio.

22. Letter and attachments from Carpenter, R. L., Uniroyal, to Berry, R., Radian Corporation. December 7, 1981. 26 p. Information on catalytic incineration at the Mishawaka, Indiana plant.

23. Martin, N. A. (Engelhard Minerals and Chemicals Corporation.) Catalytic Incineration of Low Concentration Organic Vapors. (Prepared for U. S. Environmental Protection Agency.) Washington, D.C. EPA Contract No. 68-02-3133. November 1980. 69 pp.

24. Barrett, R. E., and P. R. Sticksel. (Battelle Columbus Laboratories.) Preliminary Environmental Assessment of Afterburner Combustion Systems. (Prepared for U. S. Environmental Protection Agency.) Washington, D.C. Publication No. EPA-600/7-80-153. June 1980.

25. Letter and attachments from Dowd, E. J., ARI International, to Krohn, N. E., Radian Corporation. September 1, 1981. Information on Air Resources fluidized bed catalytic incinerators.

26. Letter and attachment from Stettenbenz, R. W., CE Air Preheater, to Krohn, N. E., Radian Corporation. September 10, 1981. Information on CE Air Preheaters emission control devices.

27. Letter and attachments from Kashmer, B. A., Johnson Matthey, Inc., to Krohn, N. E., Radian Corporation. September 8, 1981. Information on Johnson Matthey's catalytic incinerators and catalysts.

28. Letter from Madden, G. I., DuPont, to Krohn, N. E., Radian Corporation. August 17, 1981. Information on DuPont's Torvex program.

29. Letter and attachments from Rosenstein, A., Englehard, to Krohn, N. E., Radian Corporation. September 1, 1981. Information on Englehard's catalytic incinerators.

30. Letter and attachments from Janson, C., Met-Pro Corporation, to Krohn, N. E., Radian Corporation. October 9, 1981. Information on Met-Pro's emission control devices.

31. Hardison, L. C., et al. Emission Control via Fluidized Bed Oxidation. Chemical Engineering Progress. 73:31-35. August 1977.

32. Telecon. Madden, G. I., DuPont, with Krohn, N. E., Radian Corporation. September 14, 1981. Conversation about developments in catalytic incineration technology.

33. U. S. Environmental Protection Agency. Control Techniques for Volatile Organic Emissions from Stationary Sources. Research Triangle Park, N.C. Publication No. EPA-450/2-78-022. May 1978.

34. Reference 3, p. 290.

35. Telecon. Orem, S., Industrial Gas Cleaning Institute with Krohn, N. E., Radian Corporation. August 20, 1981. Conversation about number of catalytic incinerators in use.

36. Reference 1, p. 531.

37. Vincent, E. J., and W. M. Vatuvuk (EPA:OAQPS). Control of Volatile Organic Emissions from Existing Stationary Sources. Volume 8: Graphic Arts - Rotogravure and Flexography. (Prepared for U. S. Environmental Protection Agency.) Research Triangle Park, N.C. Publication No. EPA-450/2-78-033. December 1978. p. 2-6.

38. U. S. Environmental Protection Agency. Control of Volatile Organic Emissions from Existing Stationary Sources. Volume 2: Surface Coating of Cans, Coils, Paper, Fabric, Automobiles and Light-Duty Trucks. Office of Air Quality Planning and Standards. Research Triangle Park, N.C. Publication No. EPA-450/2-77-008. May 1977.

39. Frost and Sullivan, Inc. The Industrial Coatings Market. New York, 1977. p. 211.

40. Danielson, J. A., ed. (Los Angeles County Air Pollution Control District). Air Pollution Engineering Manual (Second Edition). (Prepared for U. S. Environmental Protection Agency.) Research Triangle Park, N.C. Publication No. AP-40. May 1973. p. 861.

41. Reference 40, p. 863.

42. U. S. Environmental Protection Agency. Beverage Can Surface Coating Industry - Background Information for Proposed Standards. Office of Air Quality Planning and Standards. Research Triangle Park, N.C. Publication No. EPA-450/3-80-036a. September 1980. p. 4-11.

43. Reference 42, p. 4-2.

44. Reference 40, p. 714.

45. Reference 1, p. 534.

46. U. S. Environmental Protection Agency. Metal Coil Surface Coating Industry - Background Information for Proposed Standards. Office of Air Quality Planning and Standards. Research Triangle Park, N.C. Publication No. EPA-450/3-80-035a. October 1980. p. 3-7.

47. Reference 46, p. 3-5.

48. Reference 1, p. 535.

49. U. S. Environmental Protection Agency. Automobile and Light-Duty Truck Surface Coating Operations - Background Information Document for Proposed Standards. Office of Air Quality Planning and Standards. Research Triangle Park, N.C. Publication No. EPA-450/3-79-030. October 1979. p. 4-27.

50. Reference 49, p. 4-43.

51. U. S. Environmental Protection Agency. Surface Coating of Metal Furniture - Background Information for Proposed Standards. Office of Air Quality Planning and Standards. Publication No. EPA-450/3-80-007a. September 1980.

52. Reference 51, p. 5-2.

53. U. S. Environmental Protection Agency. Control of Volatile Organic Emissions from Existing Stationary Sources. Volume 4: Surface Coating for Insulation of Magnet Wire. Office of Air Quality Planning and Standards. Research Triangle Park, N.C. Publication No. EPA-450/2-77-033. Decmeber 1977. p. 1-5.

54. Reference 53, p. 2-3.

55. Reference 3, p. 2-2.

56. Reference 38, p. 4-2.

57. Reference 28, p. 4-8.

58. Reference 38, p. 9-11.

59. Reference 38, p. 4-18.

60. Reference 38, p. 5-3.

61. U. S. Environmental Protection Agency. Pressure Sensitive Tape and Label Surface Coating Operations - Background Information for Proposed Standards. Office of Air, Noise, and Radiation. Research Triangle Park, N.C. Publication No. EPA-450/3-80-003a. August 1980. p. 3-5.

62. Reference 61, p. 3-15.

63. Reference 61, p. 3-14.

64. Reference 61, pp. 6-9, 6-10.

65. Reference 37, pp. 2-2, 2-3.

66. Anguin, M. T., and S. Anderson. (Acurex Corporation.) Acrylonitrile Plant Air Pollution Control. (Prepared for U. S. Environmental Protection Agency.) Research Triangle Park, N.C. Publication No. EPA-600/2-79-048. February 1979. p. 3-13.

67. Memo from Berry, R. S., Radian Corporation, to file October 23, 1981. Waste Stream Parameters.

68. Lovell, J. R., et al. (IT Enviroscience, Inc.) Organic Chemical Manufacturing. Volume 9: Selected Processes. (Prepared for U. S. Environmental Protection Agency.) Research Triangle Park, N.C. Publication No. EPA-450/3-80-028d. December 1980. p. III-1.

69. Reference 68, p. IV-2.

70. Reference 68, p. V-1.

71. Reference 68, p. V-5.

72. Reference 68, p. 3-1.

73. Reference 66, p. IV-3.

74. Reference 66, p. 3-20.

75. Reference 68, p. IV-7.

76. Energy and Environmental Analysis, Inc. Air Oxidation Process in the Synthetic Organic Chemical Manufacturing Industry, Background Information for Proposed Standards, Preliminary Draft. (Prepared for U. S. Environmental Protection Agency.) Research Triangle Park, N.C. EPA Publication No. 450/3-82-011A. August 1981.

77. Key, J. A., et al. (IT Enviroscience, Inc.) Organic Chemical Manufacturing. Volume 8: Selected Processes. (Prepared for U. S. Environmental Protection Agency.) Research Triangle Park, N.C. Publication No. EPA-450/3-80-028c. p. III-1.

78. Reference 77, p. V-3.

79. Reference 1, p. 530.

80. U. S. Environmental Protection Agency. Control of Volatile Organic Compound Emissions from Air Oxidation Processes in Synthetic Organic Chemicals Manufacturing Industry. Research Triangle Park, N.C. EPA Contract No. 68-02-3168. August 1981. p. 3-25.

81. Reference 76, p. 5-13.

82. Bruce, W. D., et al. Organic Chemical Manufacturing. Volume 6: Selected Processes. (Prepared for U. S. Environmental Protection Agency.) Research Triangle Park, N.C. Publication No. EPA-450/3-80-028a. December 1980. p. 14-5.

83. Reference 82, p. IV-3.

84. Reference 82, p. V-3.

85. Reference 82, p. IV-3.

86. Hobbs, F. D., et al. (IT Enviroscience, Inc.) Organic Chemical Manufacturing. Volume 7: Selected Processes. (Prepared for U. S. Environmental Protection Agency.) Research Triangle Park, N.C. Publication No. EPA-450/3-80-028b.

87. Reference 86, p. IV-6.

88. Reference 77, p. 314.

89. Telecon. Wyvill, C., Georgia Tech., with Radian Corporation. October 1, 1981.

90. Reference 1, p. 536.

91. Reference 1, p. 537.

92. Reference 1, p. 539.

93. Reference 3, p. 91.

94. Vatavich, W. M., and R. B. Neveril, Part II: Factors for Estimating Capital and Operating Costs. Chemical Engineering. 87:157-162. No. 23. November 1980.

4. Comparison of Catalytic Incineration and Alternative VOC Control Techniques

In addition to catalytic incineration, a variety of alternative VOC control technologies are available. For most VOC emitting processes, several technologies are applicable. The decision as to which is the most suitable requires a comparison on a number of technical and cost bases. This chapter provides a generalized comparison. Because surface coating processes comprise the vast majority of VOC emissions and VOC emission sources (see Section 3.3), this comparison is slanted towards this type of application.

The evaluation criteria presented earlier in Chapter 2 provide the framework for the comparison. For purposes of this chapter, the criteria are loosely subdivided into two major categories: (1) technical criteria, and (2) cost criteria. The first category concerns the overall performance of the control technique, the developmental status, application constraints, and operation and maintenance requirements. The second category concerns the overall capital and annualized costs.

The chapter begins with the identification and characterization of VOC control technologies which compete with catalytic incineration. Following this, comparisons are made on both technical and cost bases.

4.1. IDENTIFICATION AND CHARACTERIZATION OF ALTERNATIVE TECHNOLOGIES

In addition to catalytic incineration, a number of alternate control techniques may be used to reduce VOC emissions. These include the following:

- carbon adsorption
- absorption
- condensation
- thermal incineration
- process modifications

All of these techniques, with the exception of process modifications, are add-on control devices. For common surface coating operations use of a VOC capturing device and ducting to direct the VOC emissions into the control device is often required. Carbon adsorption, absorption and condensation allow the VOC's to be recovered for reuse in the process. Catalytic incineration and thermal incineration are VOC destruction devices. Process modifications, such as substituting a waterborne ink for a solvent-based ink, lower VOC emissions by reducing the amount of VOC's used in the process.

The following discussion examines each of these technologies in turn while Chapter 3 describes catalytic incineration. As in Chapter 3, application of each technology to the more common surface coating processes will receive primary emphasis.

4.1.1 Carbon Adsorption

Carbon adsorption removes VOC emissions from a waste gas stream by adsorption of the VOC's on the surface of activated carbon. The VOC's are subsequently desorbed from the carbon surface and recovered. Carbon adsorption is a mature technology that has been used for VOC emission control in a wide range of industrial processes. Its theory and principles have been exhaustively covered in the literature.[1,2] Therefore only a brief discussion of the operation of carbon adsorption units will be presented here.

There are two basic designs of carbon adsorbers: fixed-bed and fluidized-bed. Although both designs involve the same adsorption principles, there are several major differences in their design and operation. In order to illustrate these differences, fixed-bed and fluidized-bed carbon adsorption will be discussed separately.

Fixed-Bed Carbon Adsorption

Fixed-bed carbon adsorption systems contain a minimum of two adsorbers. Some larger plants have carbon adsorption systems with more than ten adsorbers. The adsorbers are cycled in and out of service resulting in each adsorber operating in a batch mode. The modes are characterized by an adsorption cycle and a desorption or regeneration cycle. This discussion of the operation of fixed-bed carbon adsorber will describe these cycles separately (See Figure 4-1).

In the adsorption cycle, the VOC gas stream is directed to an adsorber containing freshly regenerated carbon. The VOC's are adsorbed onto the surface of the carbon, and the gas stream exits with a low VOC concentration. As the capacity of the carbon bed to hold VOC's is depleted, the exit VOC concentration begins to rise. This is called the breakthrough point and indicates that the carbon is nearing saturation. Before the exit VOC concentration reaches an unacceptable level, the flow of the VOC waste gas is transferred to a freshly regenerated bed and the used bed is regenerated. Because of this batch mode operation, every fixed bed carbon adsorption system requires at least two beds.

The regeneration cycle is designed to desorb most of the retained VOC and recondition the carbon bed. There are two basic types of bed regeneration: thermal and low pressure/vacuum regeneration.

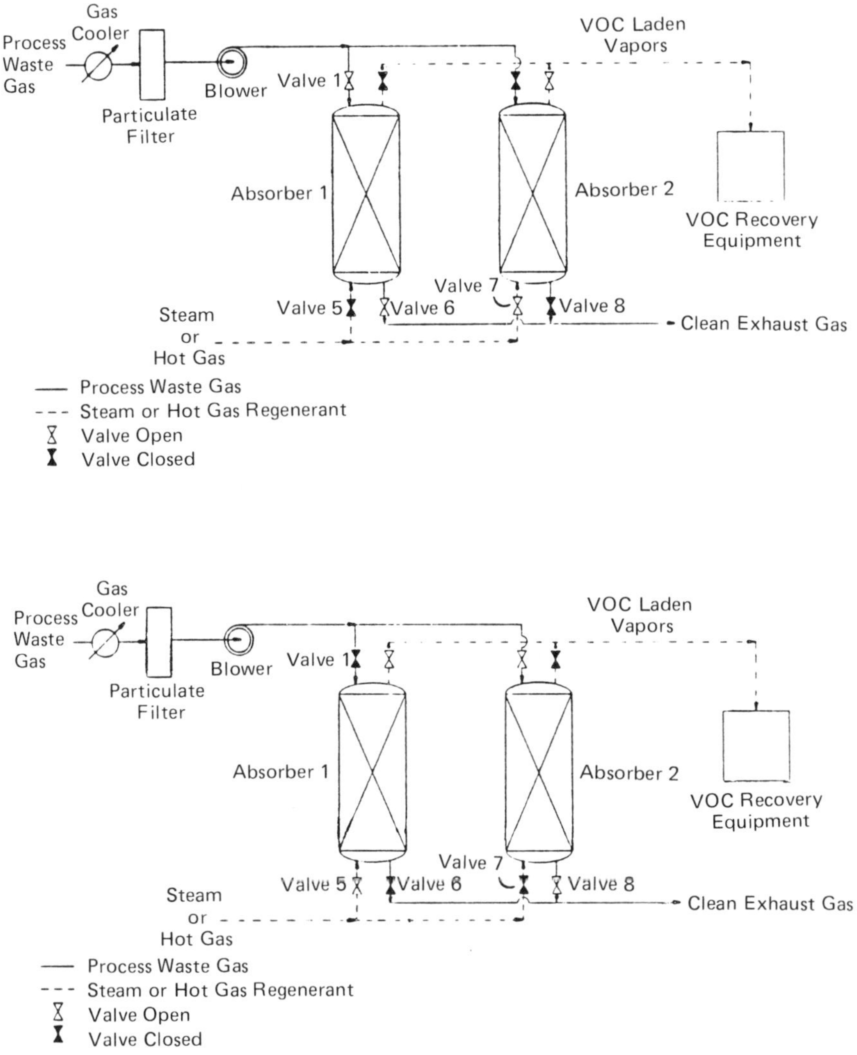

Figure 4-1. Fixed-Bed Carbon Adsorption System.

Thermal regeneration may be accomplished using steam or other hot gas such as air. Steam regeneration is accomplished by admitting low pressure steam through the bed, usually countercurrent to the VOC waste gas flow (see Figure 4-1). The steam heats the bed and causes the VOC's to desorb from the carbon. The steam and VOC mixture is then routed to a recovery system. In the recovery system the steam/VOC mixture is condensed prior to separation.

The total regeneration cycle for vacuum regeneration is similar to that for steam regeneration. In the case of vacuum regeneration, the carbon adsorber vessels are usually constructed with an external jacket. During the regeneration cycle, steam is directed through the vessel jackets to heat the carbon indirectly. The carbon is heated to at least 50° to 60°F above the VOC adsorption temperature. A vacuum is then applied to the hot carbon to remove the VOC's. The VOC vapors are then condensed and cooled with no further recovery treatment required.

When regenerating the carbon directly with steam, the condensed water/VOC mixture must be sent to a recovery system to separate the water from the VOC's so the VOC's can be reused. If the condensed VOC is miscible with water, a complex recovery system incorporating one or more distillation columns or liquid-liquid extractors will be needed. In Figure 4-2, the aqueous solution is sent to a distillation column. The VOC's are separated from the water and sent to a dessicator for final water removal. The water coming off the bottom of the distillation system is then either sent directly to sewer or to a on-site wastewater treatment plant prior to disposal.

An immiscible water/VOC mixture requires less treatment. In Figure 4-3, the immiscible mixture is sent to a decanter where the VOC's are removed through gravity separation. The waste water is either sent directly to sewer or to an on-site wastewater treatment plant. In some systems, the wastewater is stripped of the remaining VOC's in stripping columns. The air stream from the stripping column may be recycled to the carbon beds.

There are many fixed-bed carbon adsorption vendors, each marketing a slight different system. Usually, the vendors will guarantee the VOC removal efficiency in the construction contract. New systems may achieve 99 percent VOC removal efficiency with exit outlet concentrations less than 20 ppm VOC.[3] However, as the carbon ages, the efficiency may decrease. Modern carbon adsorbers are expected to achieve long-term average VOC removal efficiencies of 95 percent as expressed in several existing or proposed regulations.[3,4,5,6]

Fluidized-Bed Carbon Adsorption

Fluidized-bed carbon adsorption is a continuous process. Adsorption and desorption occur simultaneously in a single vertical vessel, which resembles a sieve tray distillation column (See Figure 4-4). The beaded

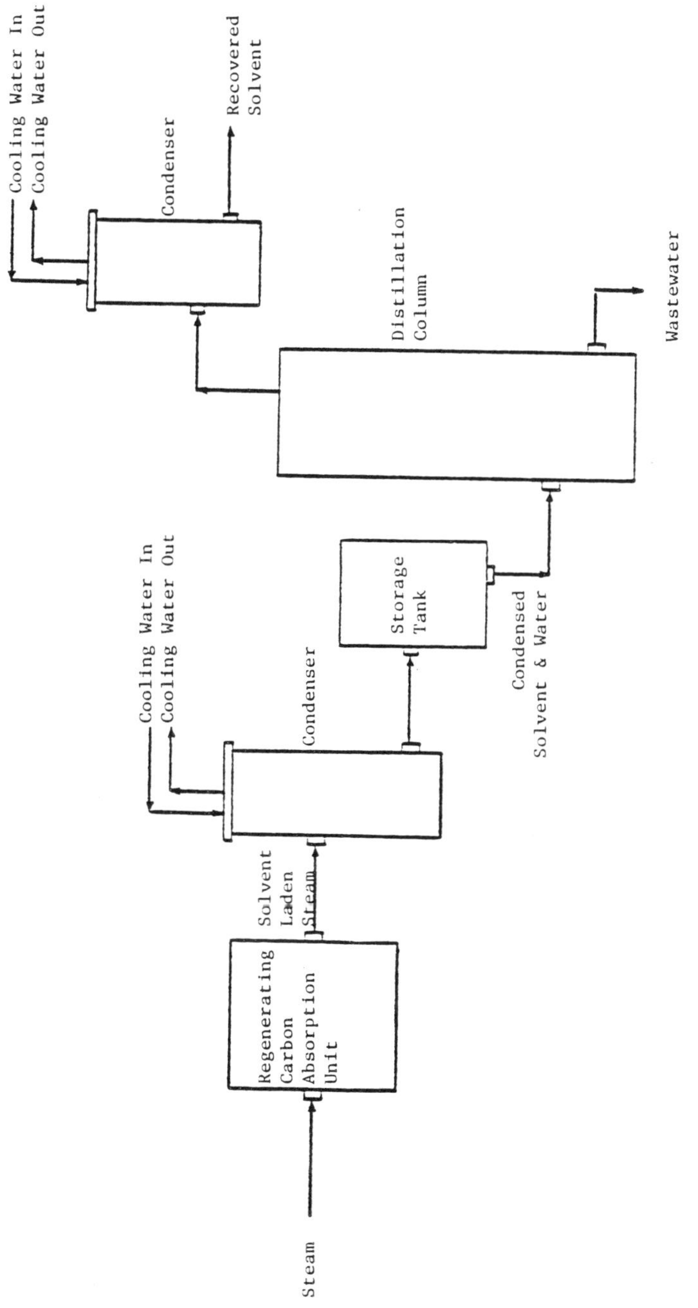

Figure 4-2. Schematic of solvent recovery by condensation and distillation.

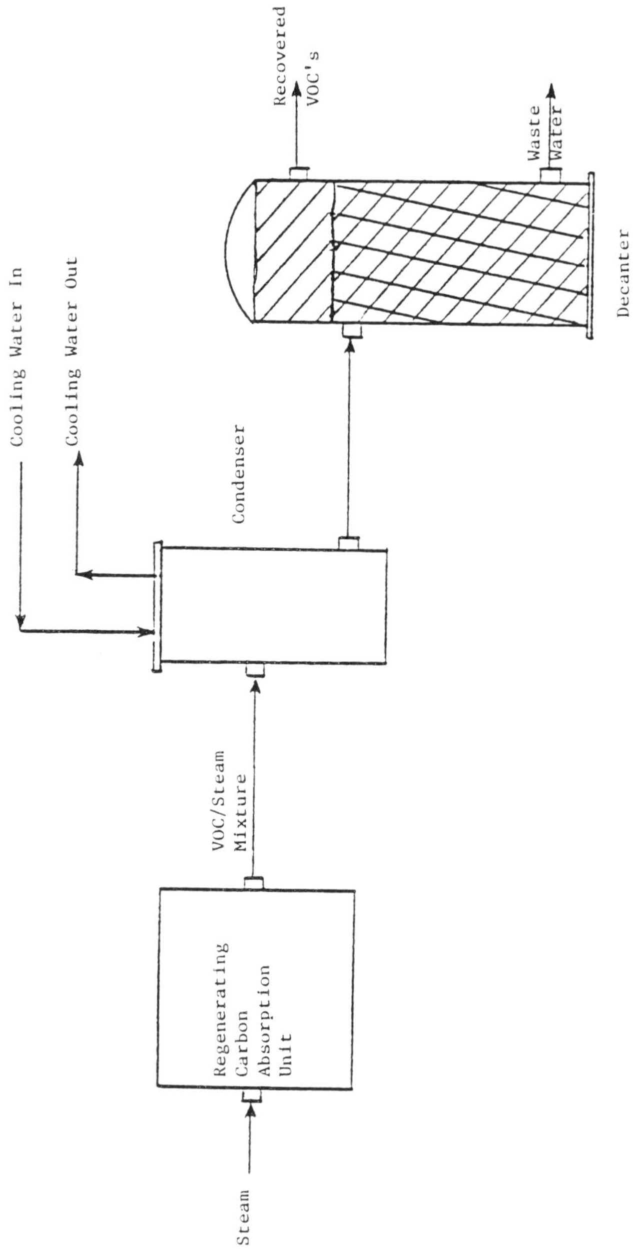

Figure 4-3. Schematic of VOC recovery by decantation.

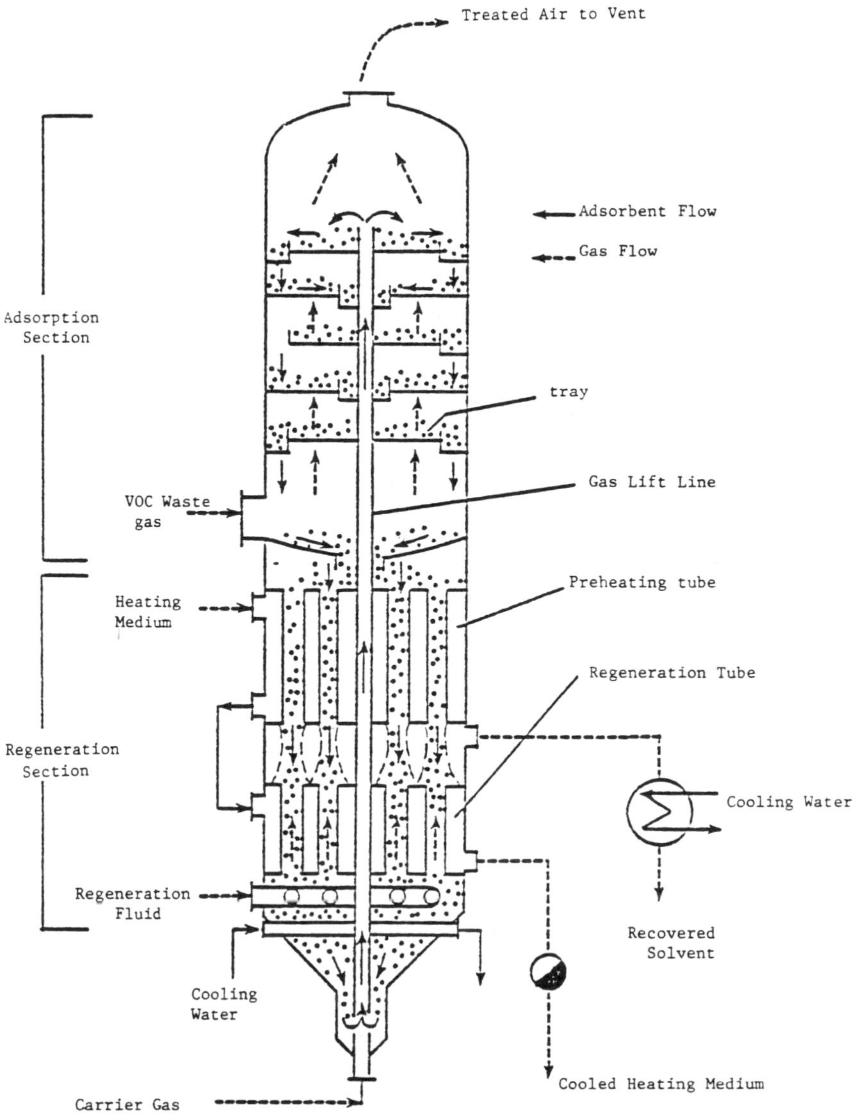

Figure 4-4. Schematic diagram of fluidized-bed carbon adsorption system.

activated carbon spheres flow countercurrent to the exhaust gas. As the VOC waste gas flows upward in the adsorber, it contacts the activated carbon in the adsorption section and the VOC's are removed. The clean gas exits through the top of the vessel while the carbon beads containing VOC's, continue down to the regeneration section. In the regeneration section, the carbon is indirectly heated with steam or other heating medium. The VOC's are then stripped from the carbon through direct contact with steam or other hot gas such as nitrogen. The regenerated carbon is then carried back to the top of the vessel by a carrier gas. The VOC vapor mixture exits through the desorption section and is condensed. When nitrogen is used, no further VOC recovery treatment is needed. If steam is used, the VOC's are recovered by decantation or distillation.

The main advantages of the fluidized-bed design arise from the single vessel design. The fluidized-bed process offers a lower energy consumption (for a given VOC waste gas flow) than a fixed-bed design. However, the capital investment for fluidized-bed carbon adsorbers is higher than fixed-bed carbon adsorption.[3]

The fluidized-bed system is a new process in the United States and previous operating experience is minimal. Fluidized-bed carbon adsorption systems are used to recover solvents in several publication rotogravure plants in Japan. They are presently marketed in the U.S. by one vendor.[3] Since it is a new process, there is very little performance data. However, one pilot operation has recorded 90 percent VOC reduction efficiency using a fluidized-bed adsorber on a stream with a low VOC concentration (100 ppm).[7]

4.1.2 Condensation

Condensation is an operation in which one or more volatile components of a vapor mixture are separated from the remaining vapor by being converted to a liquid. Condensation is accomplished by reducing the temperature of the vapor mixture until the vapor pressure of the condensible component equals its partial pressure. By further reducing the temperature of the vapor mixture, the component will condense from the vapor phase into the liquid phase. During condensation, the partial pressure of the condensible component is always equal to its vapor pressure.

Figure 4-5 shows a condensation system design currently being marketed for controlling a solvent drying oven. This system employs a nitrogen inert gas atmosphere.

The inert oven atmosphere means all oxygen has been displaced from the system. The absence of oxygen allows the oven VOC concentrations to be much higher than those allowed in a conventional drying oven system since no oxygen is present to support an explosion. Gaseous nitrogen is used to provide inert oven atmospheres and liquid nitrogen is employed as a refrigerant for condensation purposes.

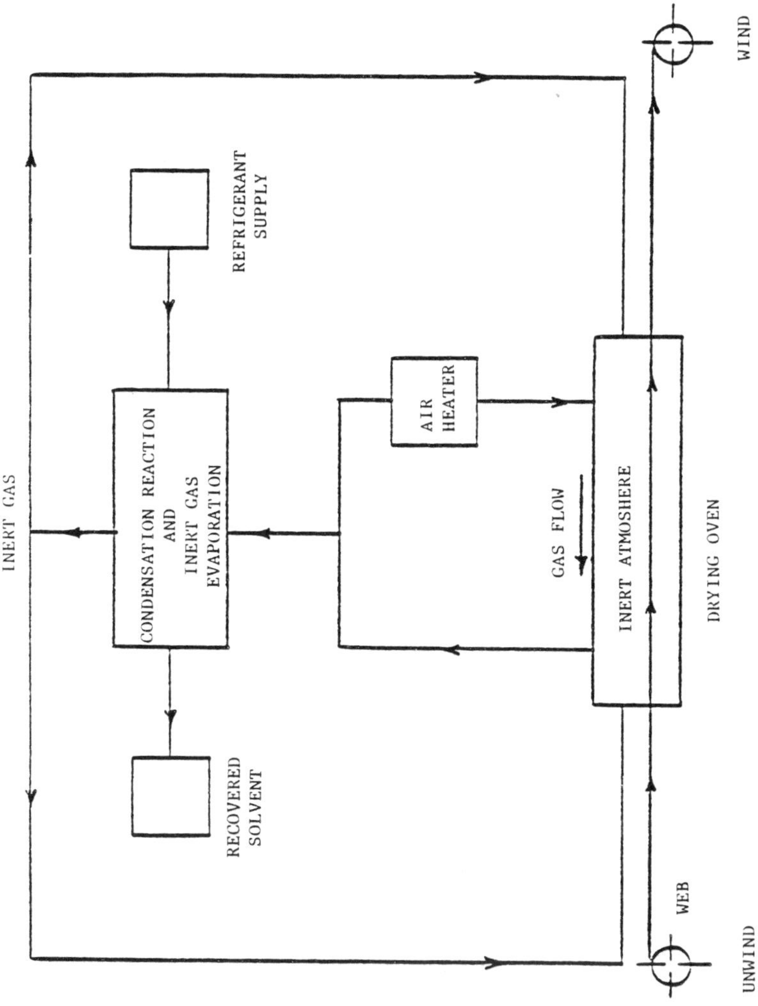

Figure 4-5. Diagram of an inert gas condenstaion solvent recovery system.

Heated, recirculated inert gases are used to dry the VOC's in the oven. Upon drying, a portion of the inert gas containing highly concentrated VOC vapors is ducted to a recovery vessel. In the vessel, the VOC's are recovered by condensation in several stages of heat exchange. If necessary, a final stage of heat exchange uses liquid nitrogen refrigeration to accomplish the desired recovery. The inert gas which has been stripped from the solvent is reused in the oven to keep the oven atmosphere balanced and inert. Inert gas vaporized from the liquid nitrogen refrigerant is also used for balancing purposes.

To successfully use inert gas condensation, the oven or other process equipment must be sealed from outside air, dust and moisture. Because of the high VOC concentrations in the inert atmosphere, air leaks could cause an explosion. Sealing is generally accomplished through the use of nitrogen gas curtains. These curtains prohibit any leakage into the system but a small quantity of nitrogen flows out of the curtained equipment. The necessary sealing process limits the application of condensation to equipment or devices such as ovens which are easily sealed.

Condensation is technically limited to sources which emit waste gases with relatively high solvent concentrations (greater than 25 percent LEL).[1] At low concentrations the condensation temperature required is infeasibly low. This constraint limits the application of condensation control to very few of the common solvent evaporation processes.

Pilot plant testing of an inert condensation system on a fabric coating facility demonstrated greater than 99 percent VOC recovery efficiency. One of the primary benefits of the inert gas condensation system for VOC control of oven emissions is energy savings. Energy is saved because large volumes of oven dilution air do not have to be heated and drawn through the oven.[8] However, since this is a new application for this technology, performance data are limited.

4.1.3 Absorption

Gas absorption is an operation in which one or more components of a vapor mixture are separated from the vapor by intimate contact with a liquid absorbent. Absorbers are designed to provide thorough contact between the contaminated gas and liquid absorbent. There are two basic designs of absorbers used for control of VOC emissions: packed and tray columns.

Packed Columns

Packed columns are vertical structures containing manufactured packing elements such as raschig rings, spiral rings, lessing rings, berl saddles and intalox saddles. The packing elements allow large surface areas and promote turbulent mixing of gas and liquid. In a countercurrent absorber the VOC waste gas stream enters the distributing space at the bottom of the column (below the packing area) and flows upward through the packing interstices. As the gas flows up through the packing, a scrubbing liquid is introduced at the top of the column and flows down over the packing. When the gas and liquid streams contact, the gaseous pollutants are absorbed from the gas into the liquid. Figure 4-6 illustrates the arrangement of a typical packed column.

Plate Columns

In contrast to packed columns, where gas and liquid absorbent are in continuous contact throughout the packed bed, plate columns employ stepwise contact. A number of trays or plates are arranged so that the gas is dispersed through a layer of liquid on each plate.

An illustration of a bubble-cap tray tower is shown in Figure 4-7. Each plate is equipped with openings (vapor risers) surmounted by bubble caps. The gas flows through the column passing through the openings in each plate and bubbles through the liquid absorbent. The liquid absorbent enters at the top of the column, and flows downward across each plate and through the down spouts. The gas bubbles through the liquid as the liquid passes down the column. The bubble caps allow the gas to achieve good contact with the liquid. The depth of liquid on the plates is maintained at a design depth which allows adequate mixing without excessive pressure loss.

Both packed and plate columns have advantages and disadvantages. Selection of one over the other depends on many factors. A comparison of packed and plate columns is contained in Table 4-1.

For common surface coating VOC emission sources, absorbers are limited to applications in which the used liquid absorbent can be reused directly without removing the absorbed VOC's. Desorption requirements are often prohibitively expensive because the residual organic concentration in the liquid must be extremely low for it to be suitable for reuse. Absorption has been used to control VOC in surface coating operations, waste handling and treatment plants, degreasing operations, asphalt batch plants, ceramic tile manufacutring plants, chromium-plating units, petroleum coker units, and varnish and resin cookers.[2] The typical removal efficiencies for absorption is from 90-99 percent.[10]

4.1.4 Thermal incineration

Thermal incineration destroys VOC emissions by thermal oxidation to carbon dioxide, water and other combustion products. In order to achieve satisfactory destruction of the VOC emissions in a thermal incinerator, three important requirements must be met.[11]

The VOC waste gas must be heated to high temperature (typically 1000°F-1600°F).

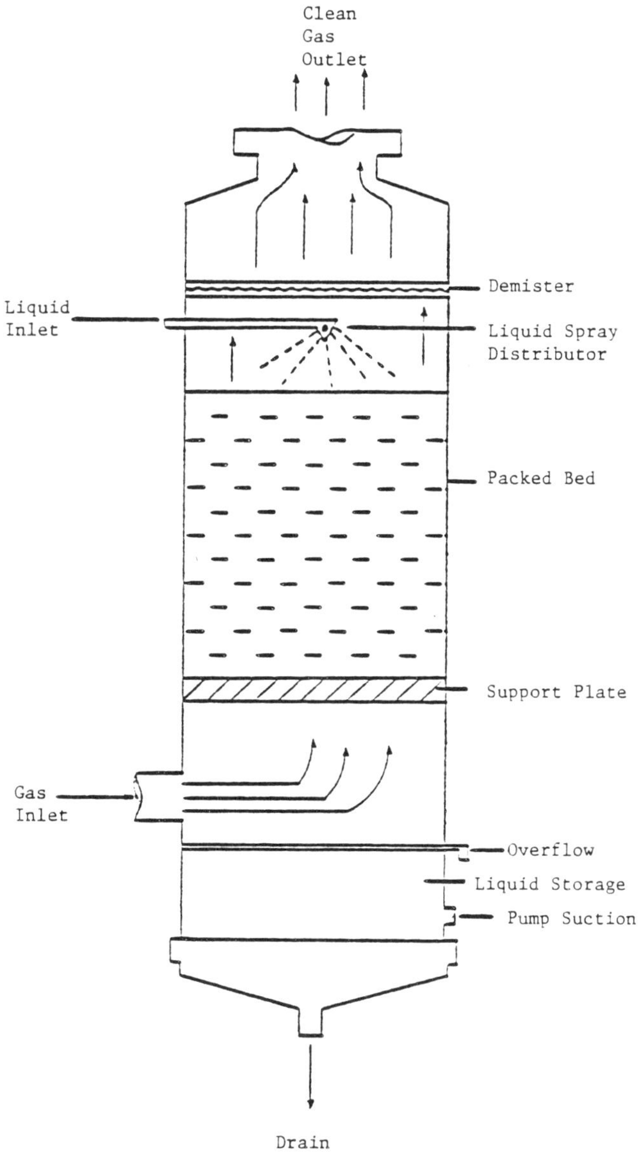

Figure 4-6. Typical packed absorption column.

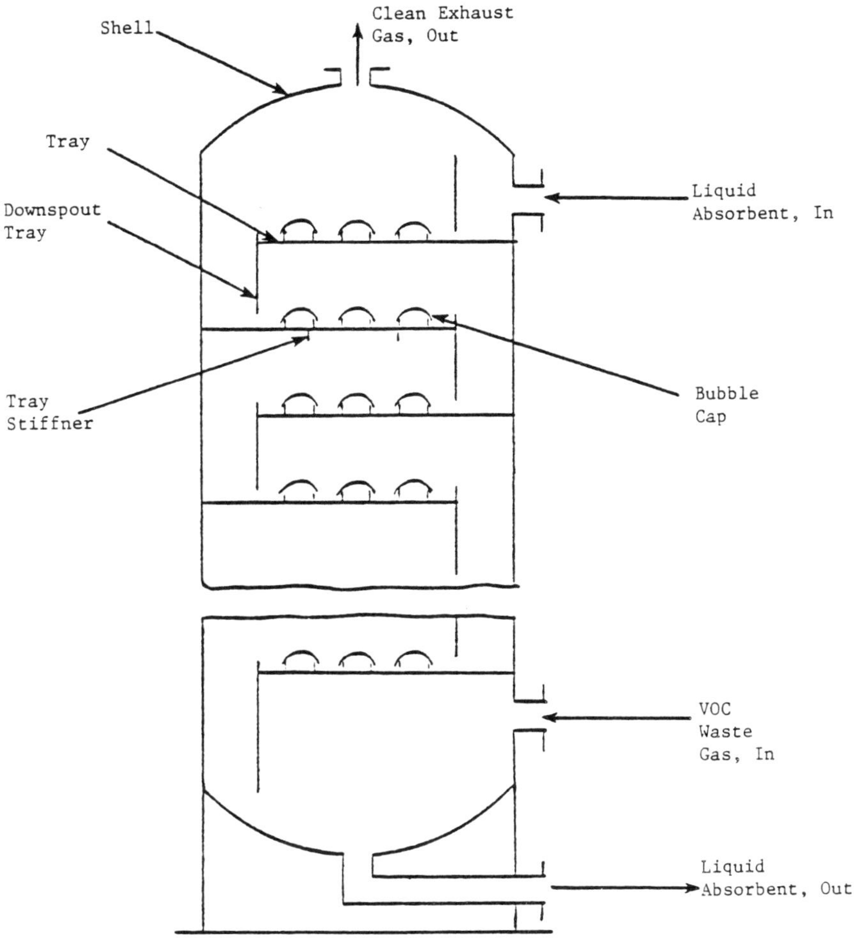

Figure 4-7. Bubble cap absorption column.

TABLE 4-1. COMPARISON OF PACKED AND PLATE COLUMNS[2]

PACKED TOWERS:

- Packed towers are less expensive than plate towers when materials of construction must be corrosion resistant.
- Packed towers have smaller pressure drops than plate towers designed for the same throughput.
- Packed towers are preferred in sizes up to 2 feet in diameter if other conditions are nearly equal.

PLATE TOWERS:

- Plate towers are preferred when the liquid contains suspended solids since plate towers are more easily cleaned.
- Plate towers are more suitable when the process involves appreciable temperature variation, since expansions and contractions due to temperature changes may crush the tower packing.

- The VOC waste gas stream must be held at this high temperature for a sufficient length of time to insure the completion of the oxidation reaction (typically 0.3 to 1 second).

- There must be sufficient turbulence within the incinerator to insure good mixing.

Most conventional thermal systems rely on large amounts of supplemental fuel to heat the VOC waste gas. In order to thoroughly heat the VOC waste gas, there must be good mixing between the fuel combustion gas and the waste gas stream. This is usually accomplished by a baffle system.[11]

A schematic of a conventional thermal incineration system is shown in Figure 4-8. The heat recovery system could be considered optional, but with rising fuel prices it is usually economical. Most recent installations include provisions for heat recovery. The primary heat recovery section is defined as the section for exchange of heat between the hot incinerator exit gas and the relatively cool VOC waste gas stream. The secondary heat recovery section is defined as the section for exchange between the hot incinerator exhaust gas and another process stream. Typical heat recovery equipment used in conventional systems are shell and tube heat exchangers, corrugated plate heat exchangers, and waste heat boilers.

A different approach to heat recovery is found in a regenerative thermal incinerator. Illustrations of a regenerative thermal incineration system are presented in Figure 4-9. The system utilizes a vertical, cylindrical combustion chamber surrounded by a series of packed stoneware beds. The VOC waste gas enters a feed header, is preheated by passing through a hot bed and then passes through the combustion chamber. The hot incinerated gases exit the chamber through a different stoneware bed transferring heat to that bed. Inlet and exhaust valves on each bed control the gas flow as the bed is depleted or saturated with heat. Each bed goes through this cooling/heating cycle every few minutes. Regenerative thermal systems can provide an energy saving over recuperative thermal systems if the VOC waste gas has a VOC content of at least 4 to 5 percent LEL and is reasonably steady state. Fluctuating VOC loadings may not allow proper interaction and timing of the heating/cooling cycles. Heat losses may increase as well since energy stored in the ceramic beds is gradually lost when the process is down.

Thermal incineration is the most widely used method to control emissions from industrial VOC sources.[12] It is capable of providing very high (greater than 98 percent) VOC destruction efficiencies when operated at very high temperatures.[11]

4.1.5 Process Modifications

This category of VOC controls includes a wide range of modifications to VOC emitting processes. In all cases, the process changes serve to reduce or eliminate the VOC emissions. Process modifications include:

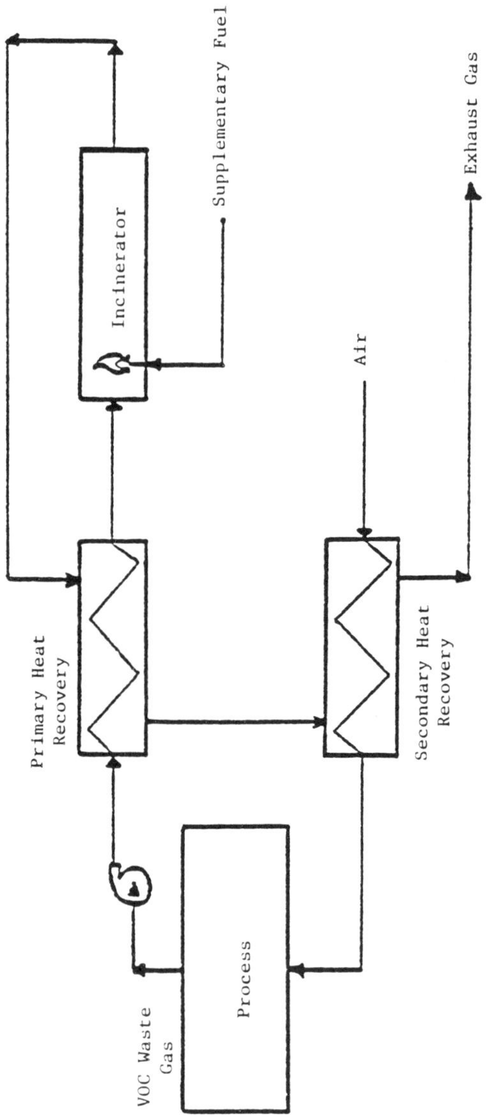

Figure 4-8. Thermal incinerator with primary and secondary heat recovery.

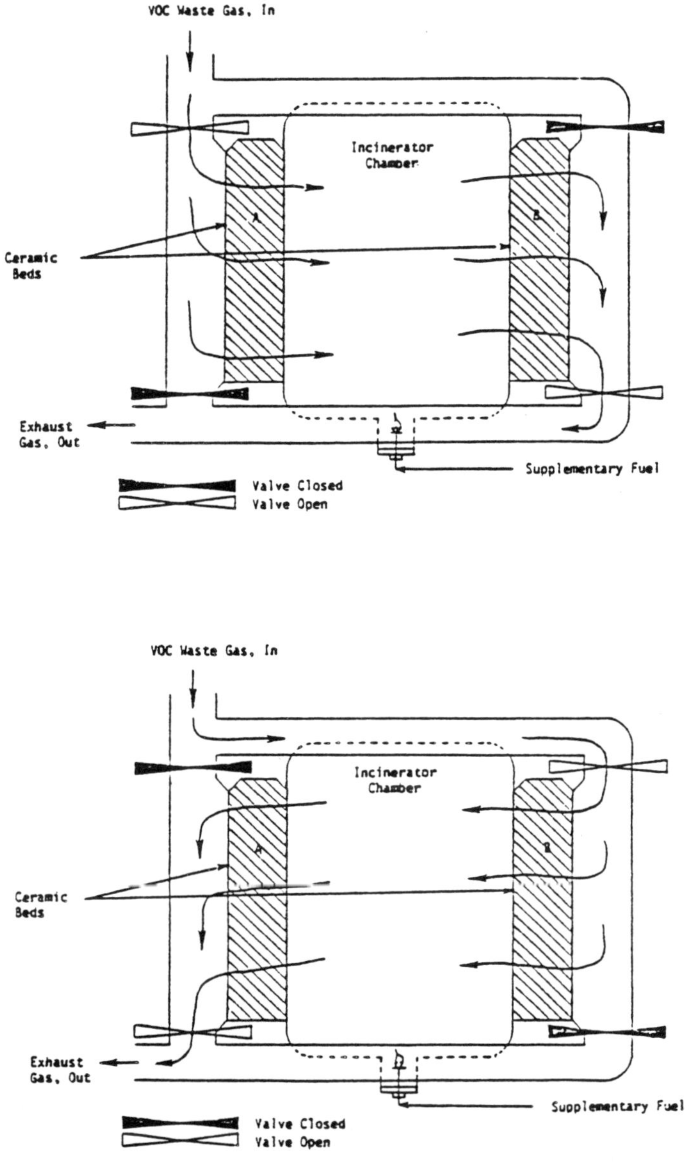

Figure 4-9. Regenerative thermal incinerator.

- changes in process conditions (i.e. temperatures, pressures, etc.),
- changes in process configuration (i.e. method of application, modes of operation, etc.),
- material substitution (i.e. use of a waterborne instead of a solvent-based coating), and
- housekeeping and maintenance procedure changes.

The applicability and performance of process modifications depends on the VOC emitting process under consideration. Each process must be examined to determine how the VOC emissions are generated. After this examination, the options available to reduce VOC's may be explored. These options may be relatively minor such as reducing the waste gas flow rate. Process modifications tend to be highly product and site specific.

For surface coating operations, material substitutions are a common process modification. For example, organic emissions can be significantly reduced by replacing conventional organic solvent-based coatings with waterborne, high solids, or powder coatings. Waterborne coatings can often be applied with the same methods used for organic solvent-borne coatings. However, waterborne coatings sometimes contain up to 20 to 30 percent organic solvent. Thus, VOC emissions are not always completely eliminated when using waterborne coatings. Product considerations may require the use of high solvent coatings, however. In many applications, product quality may not be maintained when using material substitutions or process changes.

Process modifications may reduce VOC emissions by using raw materials more effectively. For example, VOC emissions from surface coating operations can be reduced by adopting more efficient coating application methods or by changing curing techniques. Electrostatic spray coating, electron beam curing, and ultraviolet curing reduce emissions by limiting solvent contact with air.

Improved maintenance procedures and "good housekeeping" reduce VOC emissions. Vapor leaks from pumps and valves can be reduced by increasing routine maintenance and inspections. Conscientious preventive maintenance can minimize fugitive emissions from process and auxiliary equipment.[13]

4.2 TECHNICAL COMPARISON

This section presents a technical comparison of alternative VOC control technologies. Using the evaluation criteria presented in Chapter 2, the technical comparison centers on the following considerations:

- environmental and energy considerations,

- developmental status,
- applicability, and
- operation and maintenance requirements.

Another important criteria for evaluation is the overall capital and annualized cost of the control system. A separate cost analysis is presented in Section 4.3.

Of the VOC control technologies identified in Section 4.1, only the following will be examined in this section and Section 4.3:

- carbon adsorption (fixed-bed),
- catalytic incineration, and
- thermal incineration (with recuperative and regenerative heat exchange).

These technologies represent a subset of VOC controls which are generally applicable to the majority of VOC emission sources. In particular, these technologies are directly competitive for control of solvent evaporation processes which represent the vast majority of VOC emission sources. This comparison will emphasize this type of application.

The remaining technologies presented in Section 4.1 are excluded primarily because of the lack of sufficient data on the performance, maintenance and operating requirements. There are very few applications of these technologies for controlling VOC emissions. In addition, these applications are specialized to the process in which they are used. This specialization precludes a comparison of these technologies on a common basis.

4.2.1 Environmental and Energy Considerations

The environmental and energy considerations included in this discussion are the following:

- potential overall VOC removal efficiency,
- VOC removal variability,
- secondary air pollution generation,
- water pollution generation,
- solid waste generation, and
- supplemental energy use.

Table 4-2 presents a general summary of the above considerations with respect to the four control devices.

1. Potential Overall VOC Removal Efficiency

All of the VOC controls examined can potentially achieve greater than 95 percent VOC removal efficiencies with actual attainable efficiencies varying with specific application. Thermal incinerators can potentially achieve higher VOC removal efficiencies than catalytic incinerators because thermal incinerators are not subject to catalyst operating temperature constraints. Thus, thermal incinerators may be operated at very high temperatures (up to 2000°F) for maximum VOC destruction. By operating a thermal incinerator at very high temperatures, VOC's which are difficult to destruct can be oxidized to a higher degree than possible with catalytic incinerators.

However, all of these technologies are typically designed and operated to achieve lower than ultimate VOC removal efficiencies for several reasons:

- thermal incinerators would require excessively high amounts of supplemental fuel;
- catalytic incinerators would require larger catalyst beds and would also require excessive amounts of supplemental fuel;
- carbon adsorption systems would require larger carbon beds and require more energy for frequent carbon regeneration.

Thus, it is usually more economical to operate and design these systems at removal efficiencies somewhat lower than the ultimate potential of the technology. In all but the most demanding applications, all four technologies examined can provide adequate VOC removal efficiencies. In cases where extremely high removal efficiencies and/or extremely low outlet concentrations are required, thermal incineration may be preferred.

2. VOC Removal Variability

If properly designed, new carbon adsorption systems can achieve greater than 95 percent VOC removal efficiencies. There is a potential for the VOC removal efficiency to decrease as the carbon ages. This potential degradation is usually overcome by replacing the carbon every two to five years. Long- term tests on carbon adsorption systems to determine degradation in VOC removal efficiency with time are currently being conducted by the EPA.[14] If very high destruction efficiencies (i.e. greater than 98 percent) are required over time intervals of two-to-five years, thermal incineration may be preferrable to carbon adsorption controls.

As discussed in Chapter 3, if properly designed, catalytic incinerators are capable of achieving greater than 95 percent VOC removal efficiencies.[26]

TABLE 4-2. ENVIRONMENTAL AND ENERGY CONSIDERATIONS

Control Device	Potential Overall VOC Removal (%) (1)	VOC Removal Variability (2)	Secondary Air Pollution Generation (3)	Water Pollution Generation (4)	Solid Waste Pollution Generation (5)	Supplemental Energy Use (6)
Carbon Adsorption	95+	Potential for VOC removal efficiency to decrease with time. Efficiency depends on concentration.	Small amounts of particulate matter, SO_2, and NO_x from boiler used to generate steam.	VOC recovery process (i.e. decantation, distillation, extraction) create waste water containing VOC's.	Spent carbon must be either reactivated or disposed.	Depends on energy credit for recovered VOC's. Potentially very low or net credit.
Catalytic Incineration	95+	Potential for VOC removal efficiency to decrease with time.	Impurities such as nitrogen and sulfur in waste stream may cause emissions.	None.	Old catalyst support must be either reused or disposed.	Typically less than thermal incineration with recuperative heat exchange.
Thermal Incineration (Recuperative)	95+ (Potentially greater than both catalytic incineration and adsorption).	Performance generally does not degrade with time.	Impurities such as nitrogen and sulfur in waste stream may cause emissions.	None.	None.	Potential highest depending on amount of heat recovery equipment used in the system.
Thermal Incineration (Regenerative)	95+ (Potentially greater than both catalytic incineration and	Performance generally does not degrade with time.	Impurities such as nitrogen and sulfur in waste stream may cause emissions.	None.	None.	Typically less than thermal incineration with recuperative heat exchange.

There is a potential for the VOC destruction efficiency to decrease as the catalyst ages or becomes masked. Yearly catalyst cleanings are usually performed to remove any masking material and maintain the performance of the catalytic incinerator. Catalyst replacement is typically required at two-to- five year intervals. Again, the need for consistently high efficiencies (i.e. greater than 98 percent) may favor thermal incineration over catalytic incineration.

3. Secondary Air Pollution Generation

Secondary air emissions, are non-VOC emissions generated due to the operation of an air pollution control device. All incineration processes are capable of producing secondary air emissions. Combustion of halogen-containing, nitrogen-containing or sulfur-containing compounds produces secondary air emissions such as NO_x, SO_x, and acid gases. If the incinerators are not properly operated, partially oxidized compounds like carbon monoxide may be generated and emitted from incinerators. Small amounts of secondary air emissions are also produced from the generation of electric power that drives the fans used in the incinerator systems. Finally, fixation of atmospheric nitrogen may result in NO_x emissions from incinerators. Since thermal incinerators operate at higher temperatures, they generate more NO_x than do catalytic incinerators. Secondary air emissions are produced by steam generation for most carbon adsorption systems as well.

Secondary air emissions are summarized in Table 4-3. The secondary emissions are unitized on a VOC emission reduction basis. Since the fuel type used in the boilers varies widely, several alternative fuels are indicated for carbon adsorption. No data are available concerning secondary emissions from oil-fired incinerators.

TABLE 4-3. SECONDARY AIR POLLUTION DUE TO VOC CONTROL[a]

VOC Control Technology	Fuel	lb NO_x / lb VOC removed	lb SO_2 / lb VOC removed	lb particulate matter / lb VOC removed
Catalytic Incineration	Natural Gas	0.0036	negligible	negligible
Thermal Incineration	Natural Gas	0.0072	negligible	negligible
Carbon Adsorption[b]	Natural Gas	0.0010	negligible	negligible
	Residual Fuel Oil	0.0016	0.0042	0.0005
	Coal	0.0025	0.0132	0.0020

[a]Assuming no sulfur or particulate matter in waste gas stream.

[b]Produced by boiler used to generate steam.

4. Water Pollution Generation

Carbon adsorption systems using steam regeneration will produce water emissions. Depending on the type of VOC recovery process used, the waste water from the recovery process may require further onsite treatment. Catalytic and thermal incinerators, both VOC destruction devices, do not generate waste water flows. This is a significant disadvantage for carbon adsorption systems where disposal of waste water flows is constrained. However, many locales will allow the effluent stream to be sewered, thus negating this disadvantage.

5. Solid Waste Pollution Generation

Spent or used carbon is a solid waste generated by carbon adsorption systems. Usually the spent carbon is sent back to the manufacturer for reactivation. The spent catalyst from catalytic incineration is also sent back to the manufacturer for reactivation or regeneration. The catalyst manufacturer typically removes the precious metal catalyst from the catalyst support, recovers the metal and disposes of the catalyst support in a landfill. Regardless of control technique, the amount of solid material generated can be considered insignificant for the majority of applications.

6. Supplemental Energy Usage

All four control devices typically require supplemental fuel (either directly or in the form of boiler fuel used to produce steam) as well as electrical power for fans. Carbon adsorption systems allow recovery of the VOC's for reuse in the process. The energy credit for the recovered material sometimes more than compensates for the energy use of the carbon adsorption system, resulting in a net, overall energy savings. This is a significant advantage, since incineration systems typically have large fuel requirements (see Section 4.3).

Conventional thermal incinerators, which require high operating temperatures (1000°F-1600°F), are potentially the highest energy users. Catalytic incineration systems operate at much lower temperatures than thermal incinerators (800°F-1000°F) and typically require at least 25 percent less energy. High supplement fuel use is often required by both types of incinerators on coating processes because of the downtime inherent in most coating operations. During these downtimes the VOC waste gas concentration is very low. The operation of both types of incinerators during these downtime conditions is very costly because of the need to provide large amounts of supplemental fuel to the incinerator to maintain temperatures. The temperature of the incinerator cannot be allowed to drop significantly during downtimes because once the process is operating again and emitting VOC's, the incinerator must respond quickly. Thermal incinerators, especially, cannot respond quickly enough from cold status to achieve adequate VOC destruction temperatures. Catalytic incinerators are slightly more flexible and can sometimes be operated on a standby mode. The standby mode is characterized by lower operating temperatures (400-600°F typically).

The actual energy use of any thermal or catalytic incineration system depends on the required VOC destruction efficiency, the VOC content in the waste gas stream and the amount of heat recovery used in the system. In terms of heat recovery efficiency (amount of heat in exhaust gas returned to inlet or process) recuperative thermal incinerators typically achieve 30-60 percent recovery. Regenerative thermal incineration systems can potentially achieve very high heat recoveries resulting in lower fuel usage. The manufacturer of this type of system claims that 85 percent recovery is typical. In order for regenerative thermal incinerators to achieve high heat recoveries, the content of the waste gas stream must remain above 4-5 percent LEL. However, this requirement cannot usually be met on coating processes which experience significant downtime. Thus, the energy use of a regenerative thermal incinerator system may be significantly increased to the point where a regenerative system offers no advantage over a recuperative thermal incinerator system. When operating with reasonably steady loads, however, available data indicate that regenerative systems can reduce fuel use compared to recuperative thermal incinerators.

4.2.2 Developmental Status

The four VOC control technologies examined in this report (carbon adsorption, catalytic incineration, recuperative thermal incineration and regenerative thermal incineration) have been used to control VOC emissions in a wide number of industries. All four devices can usually be purchased as packaged, prepiped, prewired systems or can be custom designed. All are widely available on the commercial market.

As an overview, carbon adsorption and thermal incineration (except regenerative) have been employed as VOC controls for a somewhat longer period than catalytic incineration. However, all the examined technologies are considered available and proven. Carbon adsorption has been applied for over 30 years with at least 800 units now operating in the United States.[15] Approximately 10 vendors supply units to the market. Thermal incinerators number over 2000 with approximately 30 vendors supplying units.[16] However, regenerative thermal incinerators are a recent development (first installed in 1976) with only one vendor known to be marketing this type of system.[17] Nine vendors are known to supply catalytic incinerators systems and/or components. Over 500 have been installed with the majority of installations occurring after 1970 (see Chapter 3).

4.2.3 Applicability

This section examines the technical constraints which would limit the applicability of VOC controls. The focus will be to characterize the limitations of each technology in controlling VOC emissions from common solvent evaporation processes. Areas examined include the following:

- waste gas temperature requirements,
- susceptibility to contaminants,
- sensitivity to VOC species,
- sensitivity to varying VOC concentrations,
- size and space requirements, and
- emission data requirements.

Table 4-4 summarizes the following discussion.

1. Waste Gas Temperature Requirements

Only carbon adsorption presents a potential user with waste gas temperature constraints. Adsorbers function best with inlet gas temperatures less than 100 degrees F. For this reason many installations include a gas cooler prior to the bed to reduce gas temperatures. However,

TABLE 4-4. GENERAL APPLICABILITY OF CONTROL DEVICES

Control Device	Waste Gas Temperature Requirements (1)	Susceptibility to Contaminants (2)	Sensitivity to VOC Species (3)	Sensitivity to Varying VOC Concentrations (4)	Size and Space Requirements (5)	Emission Stream and Process Data Requirements (6)
Carbon Adsorption	Stream must usually be less than 100°F. Cooler may be required.	Carbon is susceptible to fouling by high boiling point compounds such as resins.	Streams containing multiple VOC's will be more difficult to recover. VOC's miscible with water could require complex recovery treatment (i.e. distillation).	Recovery efficiency may vary.	Size is dependent on complexity of recovery system. Typical system requires as much or more land as thermal incinerator.	Requires more extensive data than other devices.
Catalytic Incinerator	Essentially none. Very high temperature streams (greater than 900°F) are unsuitable.	Catalyst is susceptible to erosion, masking and poisoning from particulate and catalyst poisons.	Operating temperature may be adjusted to achieve desired performance.	Little or none.	Smaller than thermal units. Can be used typically for roof installation.	Requires general process information, VOC waste gas stream characteristics, (including presence of poisons and/or maskers).
Thermal Incineration (Recuperative)	None.	Corrosion can be a problem with certain contaminants.	Operating temperature may be adjusted to achieve desired performance.	Little or none.	Typically equivalent or higher than catalytic.	Minimal.
Thermal Incineration (Regenerative)	None.	Corrosion can be a problem with certain concontainments.	Operating temperature may be adjusted to performance.	Variability could affect energy use.	Typically not used for roof installation. greater than conventional thermal system.	Minimal.

this design may cause VOC condensation prior to adsorption in the carbon bed which in turn gives rise to carbon fouling problems. Filtering is one solution to this problem. In instances where cooling and/or filtering of waste gas is required, the costs of carbon adsorption versus other technologies must be carefully weighed. In general, the solvent recovered must be of relatively high value in these instances for carbon adsorption to be the least costly alternative.

2. Susceptibility to Contaminants

Thermal incineration holds a decided advantage over carbon adsorption or catalytic incineration with respect to "dirty" streams. Little quantitative data are available to characterize the types and amounts of various contaminants which can and cannot be tolerated by these two technologies. Adsorbers are susceptible to resins and high molecular weight organic binders which are difficult to desorb from the carbon bed. Catalyst beds are susceptible to these same compounds as well as catalyst poisons (see Chapter 3). It is important to emphasize that, in most cases, more frequent replacement of carbon and catalyst can alleviate these concerns. Again, costs must be considered. There are few instances where use of carbon adsorption or catalytic incineration is totally precluded.

3. Sensitivity to VOC Species

The type of VOC in the waste gas stream will influence the operating conditions for each technology. Incinerator temperatures and adsorption cycles will be adjusted to suit a particular VOC specie or combination of species. The only significant difference between the technologies is found with mixtures of VOC types. In cases where a variety of VOC's are being emitted a carbon adsorption system may be at a disadvantage. The incinerator can be adjusted, within limits, to achieve the desired destruction of the mixture. The adsorber, on the other hand, may experience difficulty with respect to desorbing capability and subsequent solvent separation. For this reason, carbon adsorbers are particularly suitable for processes in which the waste stream is reasonable consistent with respect to VOC type. In the surface coating industry, processes which apply only a limited number of coating formulations are most ameniable to control by carbon adsorption. Processes which apply a variety of coatings for different customers can typically be controlled best by incineration.

4. Sensitivity to Varying VOC Concentrations

Varying waste gas VOC loadings does not usually preclude the use of any type of control. Carbon adsorption systems tend to be constant outlet devices (the exit VOC concentration is relatively independent of the inlet VOC concentration). Thus the efficiency tends to drop as VOC concentration in the inlet drops. Incinerators tend to exhibit less efficiency drop with decreases in VOC loading and may be adjusted over short time intervals for higher temperatures.

A drop in efficiency may or may not be a problem depending on whether destruction efficiency or outlet concentration is the primary performance criteria. Some regulations do include a VOC percent reduction requirement. In instances where the VOC loading is highly variable and the VOC removal efficiency is of primary importance, incinerators would have a decided advantage over carbon adsorption systems. Thermal incinerators with recuperative heat exchangers are potentially able to follow VOC loadings more closely than other incinerator types due to their low thermal mass. The incinerator can react quickly to changes in VOC loadings and maintain a desired efficiency. Regenerative thermal incinerators will use more auxiliary fuel per mass of VOC incinerated under varying loadings as compared to steady load conditions.

5. Size and Space Requirements

Size and space requirements are an important consideration for retrofit installations. Catalytic incinerators typically require 15 square feet per 1000 scfm of capacity.[18] They are typically lighter and smaller than the other VOC control devices. Because of this, it is feasible in many cases to install catalytic incinerators in an elevated position over the process or on the roof. Thermal incinerators are usually larger and heavier than catalytic incinerators. Carbon adsorption and regenerative thermal incinerators are typically larger and heavier systems than recuperative thermal or catalytic incinerators. Although roof installation is sometimes feasible, it is not common for these two control systems. For retrofit installations where land area is at a premium, catalytic incinerators have an advantage.

6. Emission Stream and Process Data Requirements

A final consideration is the amount and type of emission data required for each type of control. Complete emission stream characterization is desirable for any VOC control device, however, minimum requirements do vary with the system. Carbon adsorbers and catalytic incinerators require the most data. Application of both technologies require knowledge of potential masking, fouling, and poisoning agents. In addition adsorber applications typically require data concerning the types of VOC in the waste gas stream in order to design separation systems. To reiterate, this requirement tends to limit adsorber applications to those instances where the operator has direct control over the type and magnitude of VOC emissions generated. For surface coating processes this generally requires the operator to know the type of solvents used in the applied coatings.

Thermal incinerators also require a general characterization of the emission stream. This characterization assists the proper design of the unit (i.e. combustion chamber size, burner configuration, heat recovery system, etc.) and also targets initial operating conditions such as incinerator temperature. However, detailed knowledge of emission characteristics is seldom required especially if the process has been successfully controlled in similar installations. After startup, the operation is usually optimized by adjustment in operating temperature until the desired performance is achieved. This flexibility in operation is important for industrial plants where the emissions are uncharacterized, subject to change, variable over time, etc. For example, a large chemical plant complex may operate a central thermal incinerator which accepts VOC emissions from numerous points scattered about the plant site. The operational flexibility afforded by such a system assures adequate performance over a wide range of emission loadings.

4.2.4 Operation and Maintenance Requirements

Operation and maintenance (O&M) considerations include a variety of concerns relating to the day to day operation of a control device. Like many other technical considerations, the O&M required in any instance will depend on site specific factors. These factors will include the following:

- severity of service (i.e., Does the stream give rise to corrosive acids?),
- characteristics of the emission stream (i.e., Does it contain particles, poisons, etc.?),
- availability of trained technical personnel familiar with O&M of the type of device installed,
- use of a preventive maintenance approach to O&M,
- location of equipment (i.e. outside in the weather or inside), and
- quality of original equipment and spare parts.

This discussion will focus on those aspects of O&M specifically associated with the use of different types of VOC control devices. For purposes of discussion, O&M considerations are broken down as follows:

- overall equipment life,
- overall reliability,
- operation labor requirements,
- maintenance labor and materials requirements,
- startup/shutdown procedures, and
- potential for equipment damage.

Table 4-5 summarizes the O&M comparison for each of these categories.

TABLE 4-5. OPERATION AND MAINTENANCE REQUIREMENTS

Control Device	Overall Equipment Life[a] (1)	Overall Reliability (2)	Operation Labor Requirements (3)	Maintenance Labor and Materials Requirements (4)	Startup/ Shutdown Procedures (5)	Potential for Equipment Damage (6)
Carbon Adsorption	Approximately 10 years (except for carbon beds).	Potentially lower than incineration systems due to complexity.	Potentially high especially if distillation systems used. Skilled personnel may be required.	Potentially high with replacement of carbon beds at infrequent intervals.	Fairly complex depending on installation, however, very quick startup.	Somewhat greater than incinerators, especially if stream is hot and cooling used.
Catalytic Incineration	Approximately 10 years (except for catalyst beds).	Approximately equivalent to thermal inciner- ation systems except on dirty streams.	Low.	Low, unless emission steam dirty. Cleanings or replacement of catalyst at infrequent intervals.	Simple; need to bring temperatures up at controlled rate.	Somewhat greater potential than thermal.
Thermal Incineration (Recuperative)	Approximately 10 years.	Potentially highest due to simplicity; break- down of refractory and insulation may be a problem due to high temperatures.	Low.	Low unless corrosive species present.	Simple; need to bring temperatures up at controlled rate.	Low unless corrosive species present.
Thermal Incineration (Regenerative)	Approximately 10 years.	Potentially lowest of incineration systems due to complexity.	Somewhat higher than other incineration systems, especially with fluctuating VOC loadings.	Low, however, maintenance of valves, dampers, and flappers can be high systems.	Fairly simple, however, procedures may be more complex than recuperative	Low.

1. Overall Equipment Life

Available data indicates no significant differences in equipment life for each of the four systems considered.[19] Carbon adsorption and catalytic incinerators do require replacement of beds at infrequent (2 to 5 years typically) intervals, however, this replacement is considered a maintenance requirement.

2. Overall Reliability

It is important to note that all systems have been demonstrated as effective and reliable VOC control devices. No data are available to quantitatively assess the comparative reliability of the systems examined. However, it can be noted that both the carbon adsorption and regenerative thermal incineration systems are batch mode systems in which large numbers of valves, dampers, flappers, controllers, etc. are used to switch airflows. This complexity creates the potential for more reliability problems. Another area of concern with both carbon adsorption and catalytic incineration systems involves problems with masking, fouling and corrosion in the beds. Resins contained in coatings can volatilize in ovens and enter the beds. In carbon adsorption systems the adsorber is typically operated at 100°F or less during the adsorption cycle creating the potential for condensation of these resins in the carbon bed. Ketone solvents can breakdown on the carbon to form corrosive acids and peroxides.[8] Many adsorption systems include corrosion resistant stainless steel structures and pH adjustment devices for recovered VOC mixtures. Such measures will minimize these reliability concerns.

3. Operation Labor Requirements

Operation labor includes the direct involvement of personnel at the plant site in the day to day operation of the control device. All the incineration systems require little operator involvement. Various references mention approximately 0.5 man-hours of operator labor per shift (see Section 4.3). The regenerative system can be expected to require somewhat higher operator involvement due to the greater complexity of the system; however, no data are available. Carbon adsorption systems can be expected to require greater operator attention than incineration systems. Available data indicates greater than one hour of operator involvement required per shift.[20]

4. Maintenance Labor and Material Requirements

Maintenance requirement trends among the systems considered are similar to operation requirement trends. Thermal incinerators are relatively maintenance free with most maintenance required for the burner and control system. Regular inspection of burner components and controls is required. In addition, the high temperatures associated with thermal incinerators would require more frequent inspection and replacement of refractory linings. Temperature cycling is potentially the most destructive influence. Regenerative systems require maintenance of valves and controllers also. Catalytic incineration systems would require somewhat greater maintenance since cleanings and replacements of catalyst are required in addition. However, most streams require very infrequent cleanings or replacements. Carbon adsorption systems are generally the most maintenance intensive systems due to the additional equipment involved in VOC recovery.

5. Startup/Shutdown Procedures

Startup and shutdown requirements are an important consideration for the VOC control devices examined. Most applications involve intermittent process operation, thus the startup/shutdown sequence will be used often. For incineration systems, the operating temperature of the device is a key variable in determining the time required for startup or shutdown. The rate at which temperature may be increased or decreased is limited. At too high a gradient, thermal effects will crack and spall insulation and refractory and damage the incinerator structure. Thermal incinerators can be brought up to operation temperature in as little as 5 to 15 minutes, however, peak efficiency may not be attained until 30 to 45 minutes after startup. Catalytic incinerators offer a decided advantage in that required temperatures are lower. Thus, catalytic units are usually quicker in startup. Carbon adsorption systems have more complex startup/shutdown procedures. At shutdown all beds are regenerated since leaving a carbon bed with solvent adsorbed can result in premature carbon degradation. However, the unit is immediately available at startup to operate with peak efficiency. If steam is used for regeneration and the boiler is shutdown as well, steam availability may lag adsorber availability by several hours. However, many installations keep boilers up continuously to avoid boiler startups and shutdowns and because many plants have a constant steam demand. In addition, regeneration is not usually required until several hours after startup allowing sufficient time for steam temperature and pressure to build.

6. Potential for Equipment Damage

Nearly all equipment damage concerns center around problems with catalyst and carbon beds. Process upsets can result in catalyst poisoning or con- tamination. Bed fires can occur in adsorption systems resulting from spontaneous ignition of the VOC's on the carbon surface. Fires are predominantly associated with ketone solvents and are most likely to occur after fresh carbon is added to the bed. Installation of adequate control and safety equipment can reduce the possibility of bed fires or catalyst contamination to very low levels. With incineration systems, unwanted temperature excursions can damage the unit. Relatively simple shutdown controllers are used to detect extreme temperatures and prevent damage.

In summary, all the examined devices can be considered reliable VOC controls with reasonable operation and maintenance requirements. If a shortage of trained O&M personnel is anticipated, the simpler systems such as recuperative thermal incinerators would be favored.

4.3 COST ANALYSIS FOR CATALYTIC INCINERATION AND ALTERNATIVE VOC CONTROL TECHNOLOGIES

This section presents a costing methodology for catalytic incinerators as they are applied to control VOC emissions from solvent evaporation processes in surface coating operations. It also provides a cost comparison between catalytic incineration, thermal incineration, and carbon adsorption. The catalytic incineration costing methodology estimates capital and annual costs after specifying certain parameters of the waste gas stream such as flow rate, temperature and heat value and after specifying the desired heat recovery and the incinerator's VOC destruction efficiency.

In order to compare catalytic incineration to the other two VOC control technologies, several industries in which catalytic incineration is being used or is technically feasible were analyzed. Typical process parameters were specified for each industry. Then, using the costing methodology, costs for these typical processes were generated for catalytic incineration. Costs for the other technologies were obtained using previous EPA costing studies with adjustments made to put all estimates on a common basis. A detailed discussion of the results of this comparison is included in the cost comparison section. In the section following the cost comparison, the results of previous cost studies are discussed.

To briefly summarize the results of the cost comparison, catalytic incineration is generally the most economical technology for processes with low waste gas concentrations (below 15% LEL), and carbon adsorption is generally the most economical for processes with high waste gas concentrations. Since metal coating operations typically have waste streams with low VOC concentrations, catalytic incineration is generally their most economical control technique. And since paper and fabric coating operations typically have waste streams with high VOC concentrations, carbon adsorption is their most economical control technique. However, for processes with high VOC concentrations, particularly those operated with high waste gas temperatures, thermal incineration appears to be more economical than catalytic incineration. The reasons behind these conclusions are given in Section 4.3.2.2.

This cost analysis does not examine regenerative thermal incineration systems. Available cost data are considered too limited to generate costs for the cases examined.

4.3.1 Catalytic Incinerator Costing Methodology

The following section presents a methodology for estimating capital and annualized costs for catalytic incinerators. The methodology estimates these costs as functions of the following variables:

- waste gas flow rate,
- waste gas temperature,
- VOC concentration,
- VOC destruction efficiency, and
- heat recovery efficiency.

4.3.1.1 Capital Costs.

Capital costs, as defined in this study, are the total installed costs of the control system. Capital cost data were available from equipment vendors, industry groups and various published reports. When placed on a common basis, (primarily adjusting to a common year to account for inflation) these data sets were generally comparable with one exception. One study reported costs significantly higher than the other data; this data set, therefore, was not used in deriving the total installed costs presented in this study. (See the study prepared by IT Enviroscience in Section 4.3.3).

Two approaches were considered for determining the total installed costs of catalytic incineration units: (1) use of vendor-supplied curves for total installed costs as a function of incinerator flow rate ($/scfm) and (2) use of previously published equipment costs in combination with cost factors. In the latter approach, the total installed cost is determined by applying various cost factors for installation, engineering, contingencies, etc. to the cost of the basic control equipment. This approach was selected for the current study for primarily two reasons. First, the results compare favorably with those derived by the first approach. And second, this general methodology has been used in previous cost studies for other VOC control systems, notably thermal incineration. Use of the cost factor methodology insured that costs for catalytic and thermal incineration presented in Section 4.3.2 were directly comparable. No suitable cost factor methodology was available for carbon adsorption systems; cost estimates used previous EPA studies adjusted to provide a similar basis.

The catalytic incinerator equipment costs used in this study were developed by the Industrial Gas Cleaning Institute (IGCI) which compiled data from six vendors of catalytic incinerators.[21] These equipment costs are presented in Figure 4-10. The equipment costs include a skid-mounted catalytic incinerator with piping, stub stack, catalyst, instrumentation, controls, and shop installed insulation. In the heat recovery cases, they also include the cost of the heat exchanger. However, the parameters used in sizing the heat exchanger were not documented by the industry group generating these costs. These costs were adjusted from 1976 to mid-1981 dollars using Chemical Engineering Equipment Cost Indices.

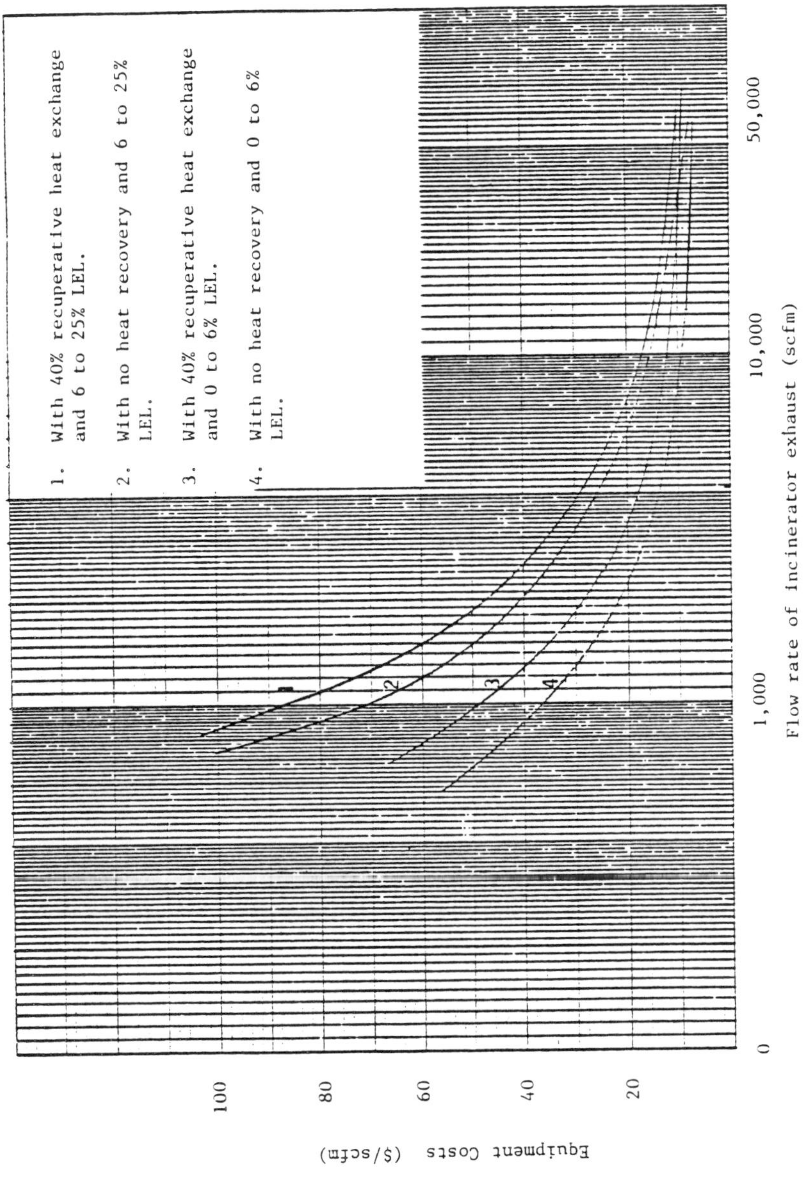

Figure 4-10. Equipment coste for catalytic incineration.

The major factors affecting the equipment costs for catalytic incineration units are the use of heat recovery, exhaust gas flow rate, and waste gas VOC concentrations. The effects of these factors can be noted in Figure 4-10. In all cases, the use of heat recovery results in an increase in total installed cost. This is due to the cost of the exchanger, the additional piping, and the additional instrumentation associated with the system. As flow rate increases, the total installed costs increase. At lower flow rates, incineration systems for low LEL waste gas streams are less expensive than those for higher LEL streams. This is the result of differing catalyst and material requirements for the different operating conditions.

It should be noted that there is a distinction between incinerator exhaust flow rate and waste gas flow rate, and the incinerator is typically designed according to the former. This is because the exhaust flow rate represents all streams that feed the incinerator. These streams include not only the waste gas stream, but also the fuel stream and the dilution air stream. Dilution air must be used in some cases so that the temperature constraints of the catalyst are not exceeded. A further discussion of this can be found in Appendix A.

As discussed earlier, total installed costs were calculated by applying cost factors to the equipment cost to yield the total installed cost. A modified cost factor method[19] was developed for catalytic incinerators and is presented in Table 4-6. Adjustments were made to the cost factors to account for the modular (skid-mounted) design of catalytic incinerators that require low site-related installation costs.

To illustrate the calculation of capital costs, an example of a beverage can coating operation will be examined. The waste gas flow is 5,000 scfm, the concentration is 10 percent LEL and 40 percent heat recovery will be assumed. From Appendix A, the exhaust gas flow rate and waste gas flow rate are found to be essentially the same (5,024 scfm) since dilution air is not required.

$$5{,}024 \text{ scfm} \times 25.5 \ \frac{\$}{\text{scfm}} \text{ (Figure 4-10)} \times 1.6 \text{ (Table 4-6)} = \$205{,}000$$

This procedure was used to determine the total installed cost for all of the representative cases examined (see Section 4.3.2).

4.3.1.2 Annualized Costs.

The total annualized cost of the control system includes direct operating and maintenance costs and annualized capital charges. The factors used in estimating the annualized costs for catalytic incinerator systems are presented in Table 4-7.

TABLE 4-6. COMPONENT CAPITAL COST FACTOR FOR CATALYTIC INCINERATORS AS A FUNCTION OF EQUIPMENT COST, Q

Purchased Equipment Costs		
Primary and auxiliary equipment (including instruments)	1.0	Q
Taxes and freight	.028	Q
	1.028	Q
Installation Costs		
Foundations and support	.082	Q
Handling and erection	.072	Q
Electrical	.041	Q
Piping/Ductwork	.021	Q
Insulation	.020	Q
Painting	.010	Q
	.246	Q
Indirect Costs		
Engineering	.051	Q
Construction and field expenses	.039	Q
Construction fee	.051	Q
Start-up	.021	Q
Performance test	.010	Q
Contingency	.154	Q
	.326	Q
Total Installed Cost Factor	1.60	Q

TABLE 4-7. BASES FOR CATALYTIC INCINERATOR ANNUALIZED COST ESTIMATES (1981)

Direct Operating Costs	
Utilities	
Electricity	$0.044/kWh[4]
Natural gas	$3.70/million Btu[4]
Labor	
Operation	$15.00/man-hour[3]
Operating hours	6,000 hours/year
Catalyst replacement	
Frequency	every 3 years[5]
Cost at 90 percent efficiency	$5.00/scfm[6]
Cost at 99 percent efficiency	$7.50/scfm[6]
Capital Charges	
Capital recovery factor	0.16275 (10 percent interest rate and 10 year life)
Taxes and Insurance	2% of total installed cost[5]
Administration	2% of total installed cost[5]
Overhead	2.5% of total installed cost[5]
Maintenance	4% of total installed cost[5]

Direct operating costs include labor, utilities, maintenance, administration, overhead, taxes, and insurance. Labor is charged at $15 per man-hour.[22] The labor requirements for catalytic incinerators have been estimated to be 0.5 man-hour per 8 hour shift.[19] The utilities needed for catalytic incinerators include electricity for fan power and natural gas to maintain the proper operating temperatures across the catalyst bed. Calculations of these utility requirements can be found in Appendix A. The electricity cost ($.044/kWh) represents the rates charged to industrial consumers in June 1981.[23] The natural gas cost ($3.70/million Btu) represents the average of rates charged to utilities and residential consumers during the same month.[23]

The catalyst used in these incinerators must be replaced periodically depending upon the process under consideration. Replacement periods typically range from several months to several years. For the purpose of this analysis, a replacement period of three (3) years was used. The cost of changing catalyst was estimated at $5.00/scfm at 90 percent VOC destruction efficiency and $7.50/scfm at 99 percent VOC destruction efficiency. For intermediate values of incinerator efficiency, these costs were linearly interpolated.

The other annualized costs (maintenance, administration, overhead, taxes, and insurance) were estimated as a factor of the total installed cost of the incinerator unit. The cost of the capital expenditure for catalytic incinerators was estimated with a capital recovery factor of 0.16275 based on 10 percent interest over the 10 year lifetime expected for the equipment.

Some of the cases considered in this study employ recuperative heat exchangers to preheat the waste gas stream before it enters the incinerator. Where heat recovery is considered, a thermal efficiency of 40 percent is assumed for the heat exchanger. (See Chapter 3 for definition of thermal efficiency). In some cases, this recovery allows the unit to operate without supplementary fuel.

It is important to note that this study examined costs of recuperative heat exchange only. In some applications other heat recovery options may be more energy efficient, (i.e. recycle systems). The actual fuel needs for these systems will depend on the individual process configuration.

An example of an annualized cost itemization is given in Table 4-8. This example demonstrates the procedure used to determine the net annualized costs for all cases presented in this study. The example in Table 4-8 is a catalytic incinerator with heat recovery used on a beverage can manufacturing line.

Figure 4-11 presents net annualized costs for catalytic incineration systems as a function of exhaust flow rate. The costs assume 95 percent VOC destruction efficiency of a waste stream containing toluene vapor at 15 percent LEL. In general, systems without heat recovery are more

TABLE 4-8. EXAMPLE OF ANNUALIZED COST ITEMIZATION FOR A CATALYTIC INCINERATOR SYSTEM[a], $1000(1981)

Direct Operating Costs		
Utilities		
Electricity	3.9	
Natural gas	14.6	
Labor	5.6	
Catalyst	10.7	
Subtotal		34.8
Capital Charges		
Capital recovery	33.4	
Taxes and Insurance	4.1	
Administration	4.1	
Overhead	5.1	
Maintenance	8.2	
Subtotal		54.9
Net Annualized Cost		89.7

[a]Example catalytic incinerator system for a beverage can manufacturing line was sized for 5,000 scfm waste gas at 10 percent LEL and 300°F. The incinerator system was designed for 95 percent VOC destruction and 40 percent heat recovery.

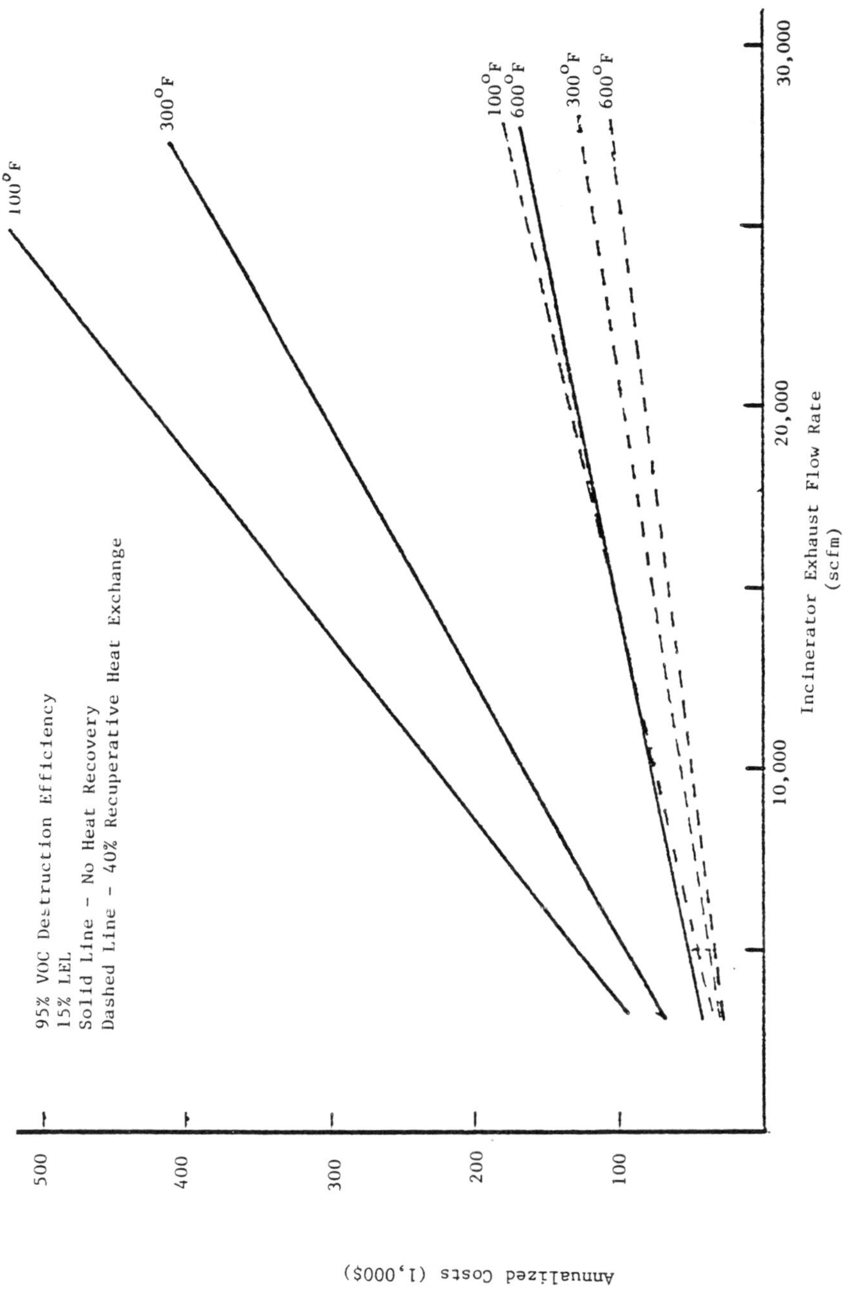

Figure 4-11. Annualized costs for catalytic incinerators for various waste gas temperatures.

expensive to operate than systems with heat recovery. The major exceptions are incinerators applied to hot gas streams with high VOC concentrations. In these cases an excess of heat is available; preheating of the inlet stream is neither desirable or economical. However, applications where this heat can be used for purposes other than preheating of the waste gas may be economical.

Figure 4-12 illustrates trends in catalytic incinerator costs for a fixed waste gas temperature (300°F) assuming 95 percent VOC destruction efficiency. As can be noted, annualized costs are considerably lower at high VOC concentrations. This does not mean, however, that catalytic incineration is the most economical VOC control technique at these high concentrations.

4.3.2 Cost Comparison for Representative Sources

This section presents a cost comparison between catalytic incineration, thermal incineration, and carbon adsorption. The comparison focuses on the estimated capital and annualized costs of applying each of these technologies to a group of "representative sources."

Two approaches were considered in developing representative sources for cost comparison. The first approach examined the range of control device and waste gas stream characteristics as noted in various applications. The representative sources consisted of all the various combinations of these characteristics. The major drawback to this approach was the unwieldy number of combinations. It was estimated that over 400 different cost estimates would be required to explore the variations in a few variables (i.e. flow rate, waste gas temperature, VOC concentration, etc.).

Instead, the approach used examines costs on an industry by industry basis. Eight sources representing seven industries were selected to cover the range of existing and potential applications. The industries examined were the beverage can, magnet wire, metal coil, vinyl coating, auto and light duty truck, rotogravure printing, and pressure sensitive tapes and labels industries. For each, a typical set of emission characteristics was developed based on the applications data presented in Section 3.3 of this report. Thus, the costs of each control device are considered representative for a typical application of the VOC technology to a particular industry. However, many applications may differ from the assumptions used in this study. In these cases, the costs may be different than those reported.

4.3.2.1 Cost Estimation Basis.

Cost estimates for catalytic incineration relied on the methodology presented earlier in Section 4.3.1. Cost estimates for thermal incineration and carbon adsorption used results of previous cost studies. Adjustments were made to bring all costs to a common basis. All estimates assumed 95% VOC destruction efficiency.

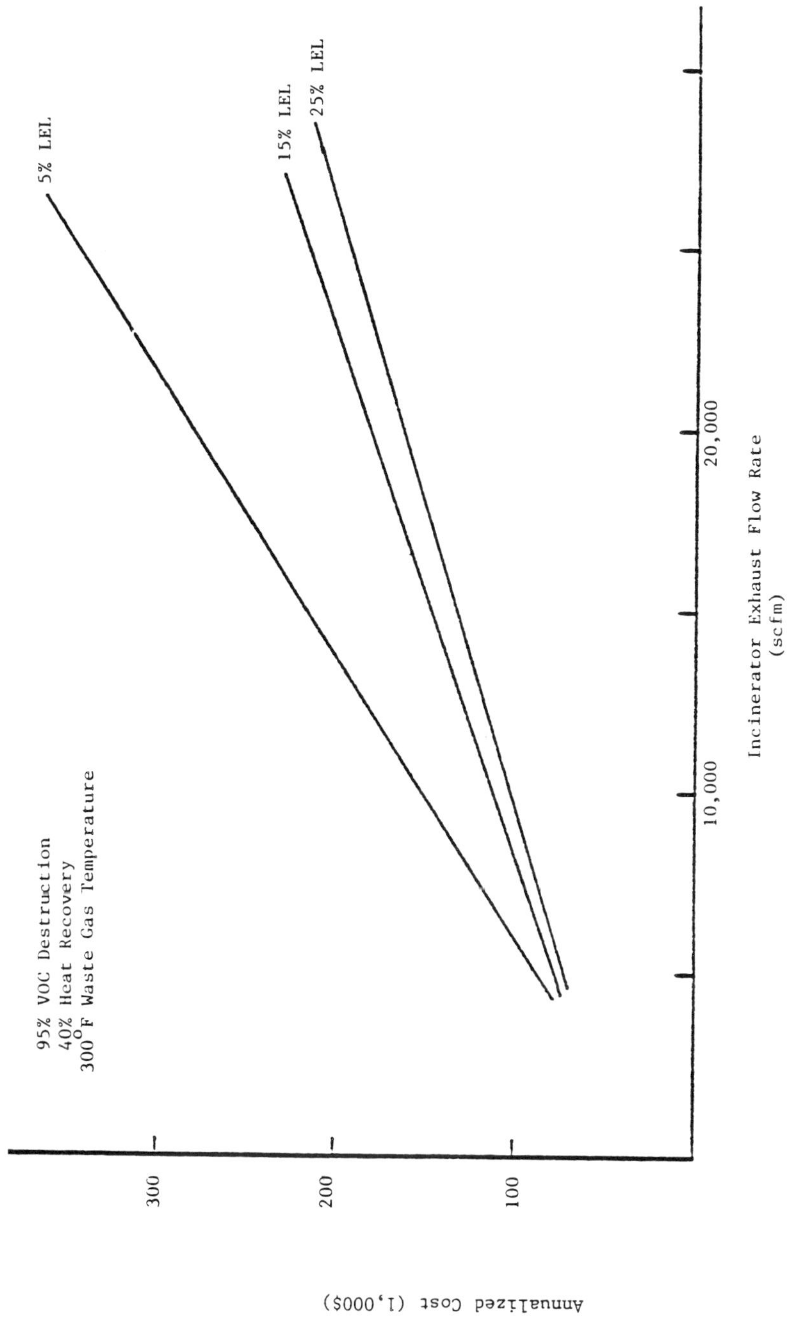

Figure 4-12. The effect of VOC concentration on annualized costs.

Thermal incinerator cost estimates used a 1976 study by the Industrial Gas Cleaning Institute.[21] This study was the major foundation of the catalytic incinerator cost methodology as well. Adjustments were made to account for changes in equipment costs, energy costs and requirements, and electrical and labor costs.

Available cost data for carbon adsorption were more limited. Costs for carbon adsorption are more variable than they are for other VOC control technologies primarily because of the cost sensitivity to the VOC species employed. Properties such as miscibility in water, corrosiveness, and recovery value will affect costs to a great degree. Other factors affecting system design are waste gas temperature and VOC concentration. Since costs for carbon adsorption are process specific, this study used industry specific costs developed by EPA in previous studies.[25] The solvents typically used in the industries for which carbon adsorption is applicable are presented in Table 4-9. Also included in this table are the current chemical prices of the solvent, the recovery values, and the type of solvent recovery processes typically used in each of these industries. Again, adjustments were required to make the results suitable for comparison.

The accuracy of the presented costs is approximately equivalent to that obtained in initial economic feasibility studies of control systems. In the judgment of the authors, all costs should be considered accurate to ±30 percent.

4.3.2.2 Cost Comparison Results.

Tables 4-10 through 4-13 and Figures 4-13 and 4-14 present capital and annual costs for catalytic incineration, thermal incineration, and carbon adsorption. The processes and the typical waste gas flow rates, temperatures, and VOC concentration were taken from Section 3.3.

Tables 4-10 and 4-12 give total capital and annualized costs for no heat recovery and 40 percent heat recovery, respectively. Tables 4-11 and 4-13 give unitized capital and annualized costs for these two cases, respectively. Capital costs were unitized according to capacity ($/scfm); annualized costs were unitized according to one million cubic feet of waste gas treated ($/$10^6$ scf). Figures 4-13 and 4-14 present the unitized annual costs of systems with no heat recovery and 40 percent heat recovery, respectively.

It should be noted that the annualized cost for carbon adsorption in some cases is negative. This simply means that the value of the recovered solvent exceeds the yearly cost of operating the system.

Using annualized costs as the basis for comparison, it is evident from Figures 4-13 and 4-14 that catalytic incineration is generally the most economical control technique for metal coating operations. These include the beverage can, magnet wire, metal coil and auto and light-duty truck

TABLE 4-9. CARBON ADSORPTION SOLVENT RECOVERY TECHNIQUES AND VALUES FOR THE APPLICABLE REPRESENTATIVE SOURCES

Industry	VOC Species and price of each ($/gal)[8,9]	Price of Solvent ($/gal)	Applicable solvent recovery technique	Recovered Value ($/gal)
Auto and Light-duty trucks	50 wt % benzene (1.50) 50 wt % hexane (1.00)	1.25	Decantation then used as boiler fuel	$.50/gal (fuel value)
Vinyl Coating	25 wt % toluene (1.50) 25 wt % MIBK (3.20) 50 wt % MEK (2.80)	2.60	Distillation	2.60
Rotogravure Printing	Solvent Naptha (1.60)	1.60	Decantation then stripping and recycling of bottom stream	1.60
Pressure Sensitive Tapes and Labels	Toluene (1.50)	1.50	Decantation	1.50

TABLE 4-10. COST COMPARISON FOR REPRESENTATIVE SOURCES WITHOUT HEAT RECOVERY

		Parameters			Capital Costs			Annual Costs (10^3\$)		
Industry	Process	Flow (scfm)	Temp. (°F)	Concentration (% LEL)	Catalytic Incineration	Thermal Incineration	Carbon Adsorption	Catalytic Incineration	Thermal Incineration	Carbon Adsorption
Beverage Can Manufacturing	Oven	5,000	300	10	170	162	a	113	169	a
Magent Wire	Oven	1,000	600	15	115	138	a	38.8	54.2	a
Metal Coil	Oven	15,000	600	15	280	209	a	118	239	a
Vinyl Coating	Oven	6,300	300	25	210	153	1,140	115	118	44.0
Auto and Light-Duty Truck	Oven Oven	2,500 15,000	300 300	5 5	69.7 223	161 209	208 811	65.1 310	119 484	158 665
Rotogravure Printing	Dryer	46,000	120	20	548	337	1,720	920	1,022	(711)
Pressure Sensitive Tape and Label	Dryer	25,000	200	25	398	231	1,000	521	419	(584)

[a]Carbon adsorption is not technically suitable for these operations. See Section 4.2 for further discussion.

TABLE 4-11. COST COMPARISON FOR REPRESENTATIVE SOURCES WITHOUT HEAT RECOVERY ON A UNITIZED BASIS

Industry	Process	Parameters: Flow (scfm)	Parameters: Temp. (°F)	Parameters: Concentration (% LEL)	Capital Costs ($/scfm): Catalytic Incineration	Capital Costs ($/scfm): Thermal Incineration	Capital Costs ($/scfm): Carbon Adsorption	Annual Costs ($/$10^6$ scf): Catalytic Incineration	Annual Costs ($/$10^6$ scf): Thermal Incineration	Annual Costs ($/$10^6$ scf): Carbon Adsorption[b]
Beverage Can Manufacturing	Oven	5,000	300	10	34.0	32.4	a	62.7	93.9	a
Magnet Wire	Oven	1,000	600	15	115	138	a	107	150	a
Metal Coil	Oven	15,000	600	15	18.7	13.9	a	22.1	46.8	a
Vinyl Coating	Oven	6,300	300	25	33.3	24.3	181	50.8	51.7	19.6
Auto and Light-Duty Truck	Oven Oven	2,500 15,000	300 300	5 5	27.9 14.9	64.4 13.9	83.2 54.0	72.6 57.7	133 89.7	175 123
Rotogravure Printing	Dryer	46,000	120	20	11.9	7.3	37.4	55.7	61.5	(42.5)
Pressure Sensitive Tape & Label	Dryer	25,000	200	25	15.9	9.2	40.0	57.9	46.3	(64.6)

[a]Carbon adsorption is not technically suitable for these operations. See Section 4.2 for further discussion.

[b]Parentheses indicate that the value for the recovered solvent exceeds the yearly operating costs.

TABLE 4-12. COST COMPARISON FOR REPRESENTATIVE SOURCES WITH 40 PERCENT HEAT RECOVERY

		Parameters			Capital Costs (10^3\$)			Annual Costs (10^3\$)		
Industry	Process	Flow (scfm)	Temp. (°F)	Concentration (% LEL)	Catalytic Incineration	Thermal Incineration	Carbon Adsorption	Catalytic Incineration	Thermal Incineration	Carbon Adsorption[b]
Beverage Can Manufacturing	Oven	5,000	300	10	205	203	a	89.7	122	a
Magnet Wire	Oven	1,000	600	15	133	181	a	44.3	57.4	a
Metal Coil	Oven	15,000	600	15	370	261	a	151	122	a
Vinyl Coating	Oven	6,300	300	25	241	209	1,140	87.3	80.3	44.0
Auto and Light-Duty Truck	Oven Oven	2,500 15,000	300 300	5 5	98.7 278	200 261	208 810	56.5 227	101 330	158 665
Rotogravure Printing	Dryer	46,000	120	20	695	475	1,720	423	449	(711)
Pressure Sensitive Tape & Label	Dryer	25,000	200	25	513	331	1,000	233	183	(584)

[a]Carbon adsorption is not technically suitable for these operations. See Section 4.2 for further discussion.

[b]Parentheses indicate that the value for the recovered solvent exceeds the yearly operating costs.

TABLE 4-13. COST COMPARISON FOR REPRESENTATIVE SOURCES WITH 40 PERCENT HEAT RECOVERY ON A UNITIZED BASIS

Industry	Process	Flow (scfm)	Parameters Temp. (°F)	Concentration (% LEL)	Capital Costs ($/scfm) Catalytic Incineration	Thermal Incineration	Carbon Adsorption	Annual Costs ($/$10^6$scf) Catalytic Incineration	Thermal Incineration	Carbon Adsorption[b]
Beverage Can Manufacturing	Oven	5,000	300	10	41.0	40.6	a	50.3	67.8	a
Magnet Wire	Oven	1,000	600	15	133	181	a	124	159	a
Metal Coil	Oven	15,000	600	15	24.7	17.4	a	28.3	22.6	a
Vinyl Coating	Oven	6,300	300	25	38.3	33.2	181	38.1	35.4	19.6
Auto and Light-Duty Truck	Oven Oven	2,500 15,000	300 300	5 5	39.5 18.5	80.0 17.4	83.2 55.0	63.2 42.1	112 61.1	175 123
Rotogravure Printing	Dryer	46,000	120	20	15.1	10.3	37.4	25.9	30.1	(42.5)
Pressure Sensitive Tape & Label	Dryer	25,000	200	25	20.5	13.2	40.0	26.1	20.3	(64.6)

[a]Carbon adsorption is not technically suitable for these operations. See Section 4.2 for further discussion.

[b]Parentheses indicate that the value for the recovered solvent exceeds the yearly operating costs.

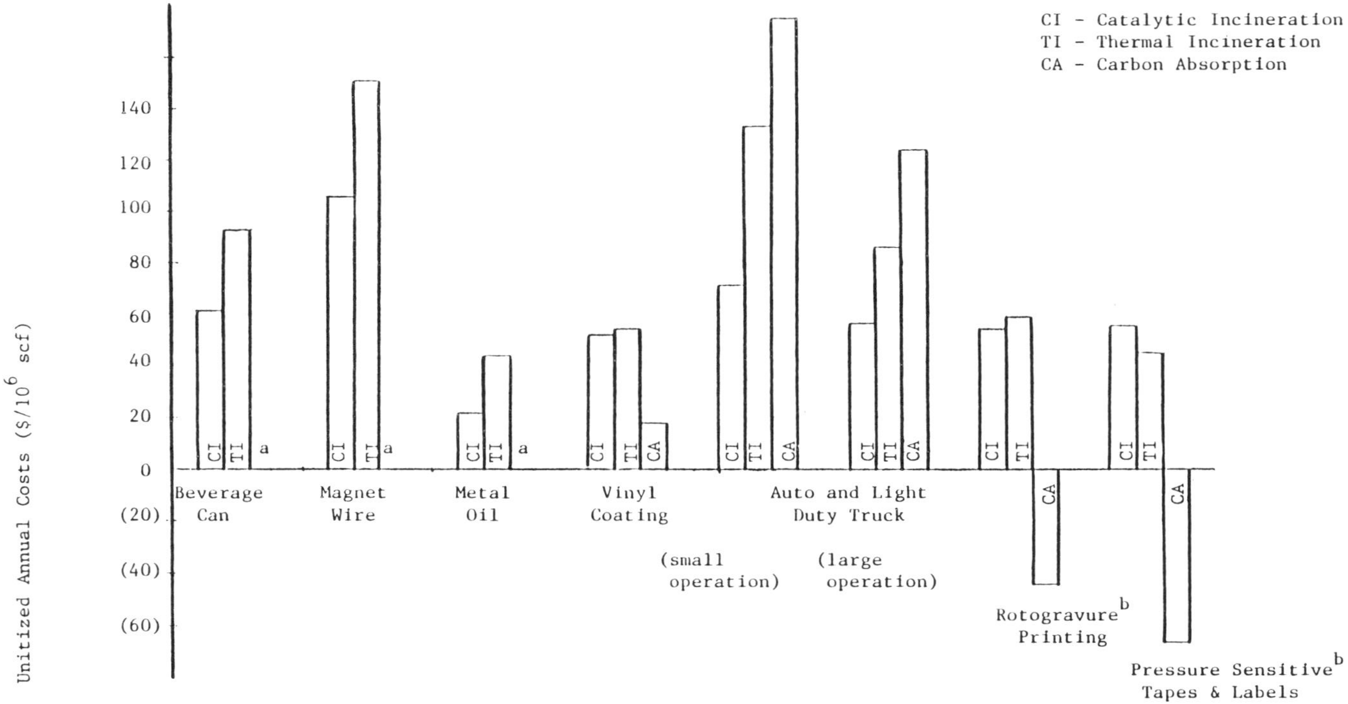

Figure 4-13. Comparison of unitized annualized costs using no heat recovery.

---

[a]Carbon absorption is not technically suitable for these operations. See Section 4.2 for further discussion.

[b]Negative values indicate that the value for the recovered solvent exceeds the yearly operation costs.

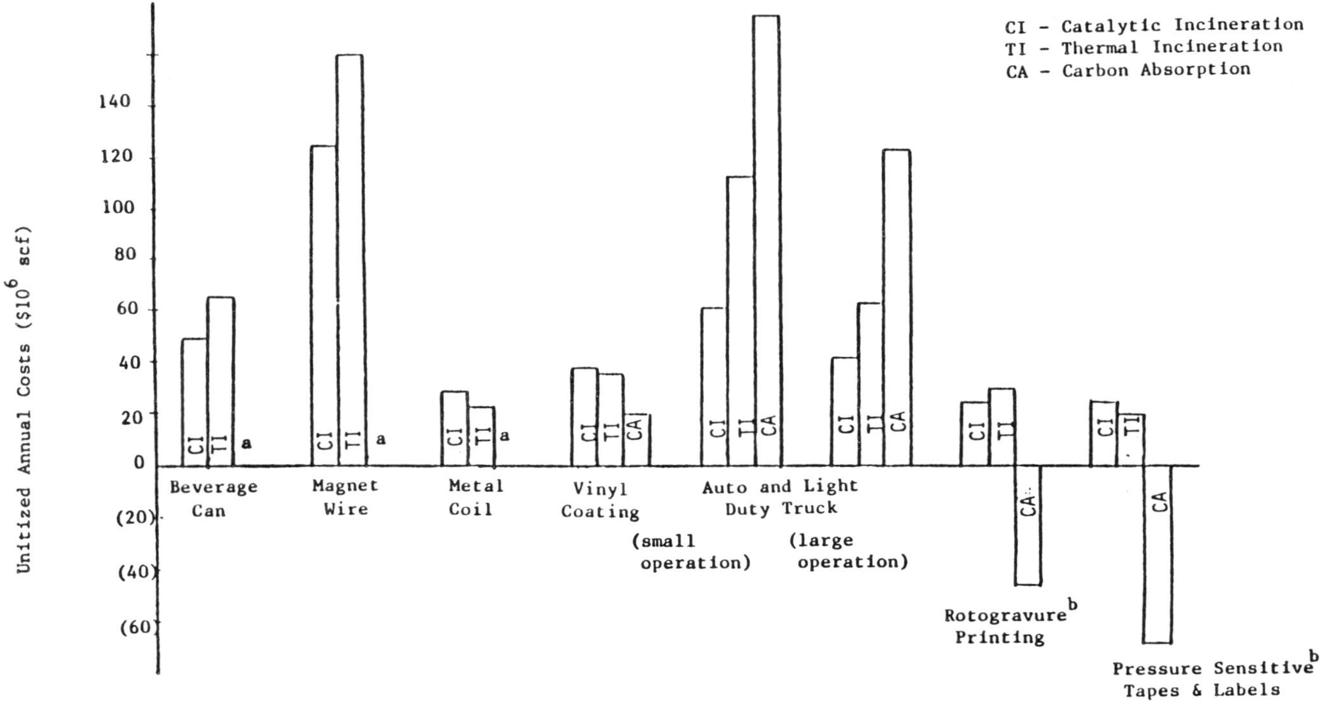

Figure 4-14. Comparison of unitized annualized costs using 40 percent heat recovery.

[a]Carbon absorption is not technically suitable for these operations. See Section 4.2 for further discussion.

[b]Negative values indicate that the values for the recovered solvent exceeds the yearly operation costs.

industries. It is also evident that carbon adsorption is the most economical control technique for fabric and paper coating operations. These include vinyl coating, rotogravure printing, and pressure sensitive tapes and labels.

The cost comparison presented in Figures 4-13 and 4-14 provide useful information for determining the most economical VOC control technique for these particular industries. Although the comparison is limited to eight processes, the results can be generalized to other processes. By relating waste gas parameters to the annualized costs of each VOC control technology for each of the eight processes, the most economical control alternative for others can be determined given their waste gas parameters.

The major factor determining the selection of the most economical control technique is VOC concentration. For low VOC concentration catalytic incineration with heat recovery is the most economical option. The exact concentration at which the other two control technologies become competitive depends on the waste gas flow rate and temperature. From the tables and figures, it appears that this concentration is approximately 15 percent LEL, or translated into fuel value, 8 Btu/scf for typical solvents. At higher concentrations, it appears that carbon adsorption is clearly the most economical alternative. For rotogravure printing and pressure sensitive tapes and labels, substantial net cost savings are achieved using this technique.

However, an important consideration for carbon adsorption systems is the coating and solvent properties. In the beverage can, metal coil, and magnet wire industries, carbon adsorption is technically unsuitable in most cases because the resins typically used in these industries condense at relatively high temperatures and foul the carbon beds. Even if carbon adsorption were suitable, there are other factors which make it prohibitively expensive. For example, the solvents used in these industries have many components that have different miscibilities in water. Because of this, the recovery value is reduced, and water treatment costs may be incurred. Many times the solvents are corrosive and require expensive materials of construction. Furthermore, the high waste gas temperatures require some type of cooling device, and if a quencher is selected, water treatment costs are increased. However, in some cases where expensive solvent mixtures such as MIBK and MEK are used, those additional capital and operating costs are offset by the solvent recovery credit. The vinyl coating industry is an example of such a case. Incineration, on the other hand, is less affected by coating and solvent properties. Therefore, it is more economical than carbon adsorption in many applications with high concentrations of relatively inexpensive solvents.

The question then is which incineration technique is more economical for waste gas streams with high VOC concentrations but for which carbon adsorption is either prohibitively expensive or technically unsuitable. Again, the choice depends on concentration. At concentrations ranging from approximately 15 percent LEL to 20 percent LEL, the annualized costs for catalytic and thermal incineration are very nearly the same. Thermal incineration becomes more competitive as flow rate and heat recovery are increased. As flow rate increases, the capital costs for thermal incineration increase less rapidly than do the capital costs for catalytic incinerators. And as the amount of heat recovered is increased, fuel costs for thermal incineration drop more dramatically than they do for catalytic incinerators. Fuel costs and capital charges are the most substantial costs affecting annualized costs.

At concentrations above 20 percent LEL, thermal incineration is nearly always less expensive because of the dilution air requirements for catalytic incinerators. Without dilution air, waste gas streams with high fuel value will produce temperatures above that which the catalysts can tolerate. These temperatures are increased even more with high waste gas temperatures. In these cases, even more dilution air must be added. The addition of air to the waste gas stream increases fuel and capital requirements for catalytic incineration. Fuel costs increase because more air has to be heated. Capital costs increase relative to those for thermal incineration because of the larger incinerator required which in turn requires more catalyst.

4.3.3 Results of Previous Cost Studies

This subsection reviews four VOC control technology cost comparisons developed in previous studies. The discussions of each will include the major process parameters, a brief description of the costing methodology, and a brief discussion of differences with and similarities to this study's methodology and results. Figures 4-15 and 4-16 show the capital and annual costs for catalytic incinerators for three of these studies (the fourth is considered confidential). All costs have all been scaled to 1981 dollars.

Report of Fuel Requirements, Capital Cost and Operating Expense For Catalytic Afterburners[21] (Prepared by IGCI).

This report was prepared in 1976 for the EPA by CE Air Preheater and the Industrial Gas Cleaning Institute. It presents capital and annual costs for both catalytic and thermal incineration systems. Four cases were examined: systems with no heat recovery, systems with 35 percent heat recovery employing recuperative exchange, systems with 55 percent heat recovery employing an additional heat exchanger, and systems with an unspecified amount of heat recovery employing recirculation of incinerator exhaust through the process. In each of these cases, annual and capital costs were examined for flow rates of 5,000, 15,000, and 30,000 scfm, inlet gas temperatures of 70°F and 300°F, and concentrations ranging from 0 percent LEL to 25 percent LEL.

The IGCI study compared the capital and annual costs for catalytic and thermal incineration. The results parallel the results obtained in the

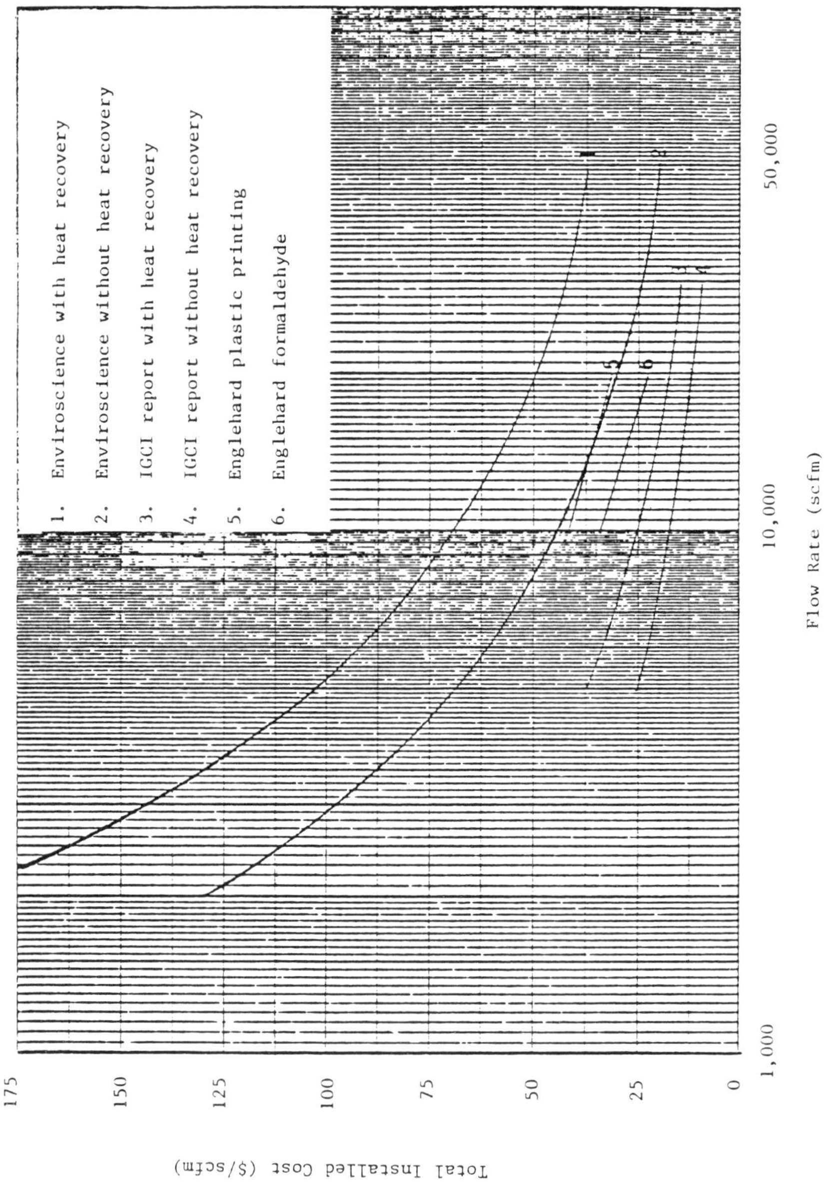

Figure 4-15. Captial costs from previous cost studies.

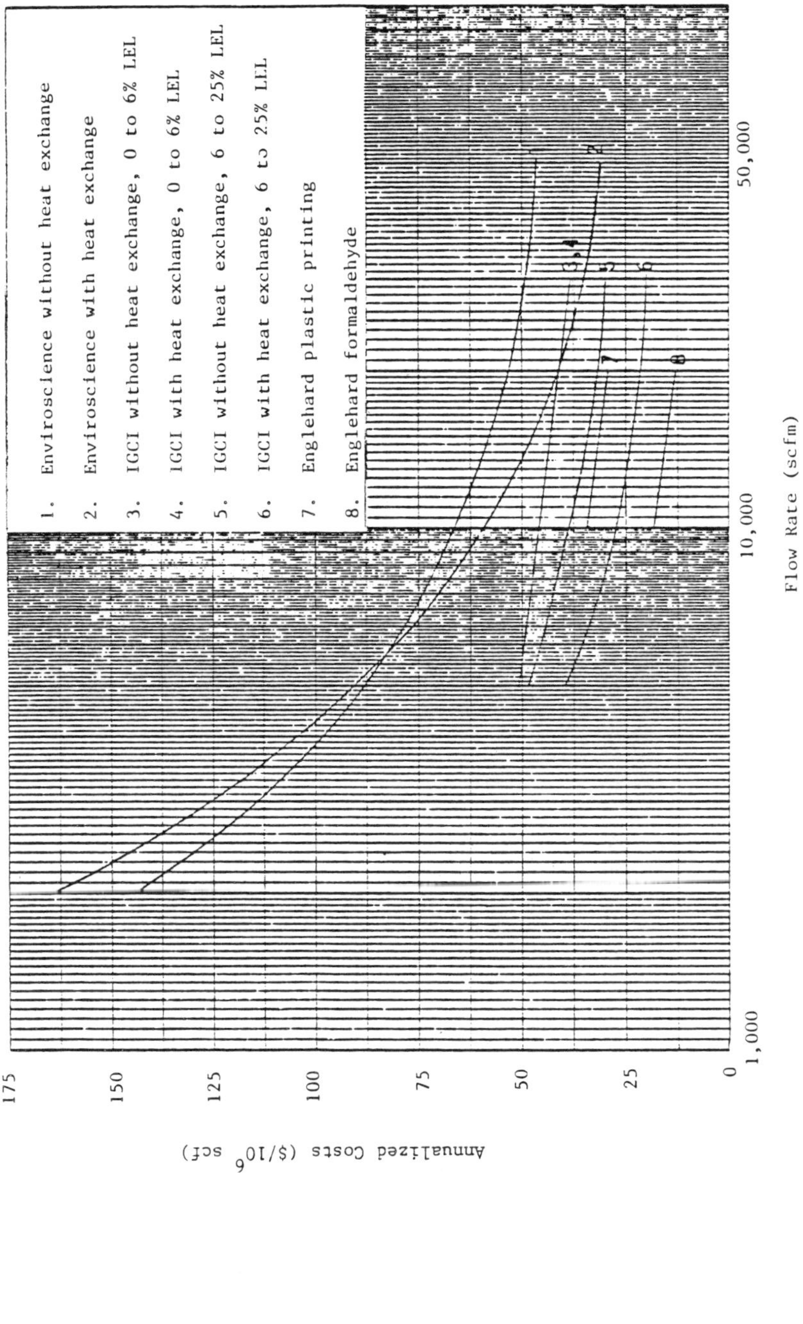

Figure 4-16. Annualized costs from previous cost studies.

study presented herein. Catalytic incineration capital costs were substantially lower than thermal incineration capital costs for low flow rates and low concentrations. For concentrations above 6 percent, the capital costs were greater for catalytic incineration than for thermal incineration. Without heat recovery, the annual costs were less for catalytic incineration, but this cost advantage for catalytic decreased with increasing concentration. However, with 35 percent heat recovery, thermal incineration annualized costs were comparable at 15 percent LEL and lower at 25 percent LEL.

Control Device Evaluation - Catalytic Oxidation[22] (Prepared by IT Enviroscience).

This report was part of a 10 volume study prepared for EPA to help support emission standards development for the synthetic organic chemical manufacturing industry (SOCMI). This report, completed in October 1980, was included in Volume 4 which evaluates the use of combustion devices to control VOC emissions.

Capital and annual costs were estimated for seven waste gas flows ranging from 700 - 100,000 scfm and for destruction efficiencies of 90 and 99 percent. No heat recovery, heat recovery with recuperative heat exchangers, and heat recovery with a waste-heat boiler were examined. The temperature of the waste gas stream was assumed to be 100°F, characteristic of absorber vent gases, and the concentrations varied from 0 to 20 Btu/scf. For four of the five concentrations considered, the waste gas contained no oxygen. Only one case was assumed to have sufficient oxygen to carry out combustion without supplemental air. This case is comparable in this respect to the representative cases examined in this report and is presented in Figures 4-15 and 4-16. The 90 percent VOC destruction efficiency case was also used in these figures.

Total installed capital costs were developed by applying certain factors to the purchase prices of the equipment. The basic pieces of equipment included catalytic oxidizers, heat recovery system, plus the catalyst, ducts, and stacks. Purchase costs were obtained from vendors and previous EPA reports.

The Enviroscience study capital cost estimates are substantially higher than the costs developed with this report's methodology. Purchase costs are approximately the same for the two studies. The difference is due primarily to differences in installation costs. For the Enviroscience study, the installation costs represent approximately three times the purchase cost, whereas for this study's methodology they represent 60 percent of the purchase costs. This difference can be attributed to the different purposes for which catalytic incinerators are installed. For many chemical processes, catalytic incineration systems are often more than a means for VOC destruction; rather they are an integral part of the process and require a custom designed and installed ducting network. On the other hand, in solvent evaporation processes, to which our costing methodology applies, the systems are installed primarily for VOC control and are usually purchased as packaged units and are skid-mounted. Freight, construction fees, engineering costs and contingencies, therefore, have a greater impact on the total installed cost for custom designed units than they do for packaged units.

Catalytic Incineration of Low Concentration Organic Vapors (Prepared by Englehard).[26]

This report was prepared for IERL by Englehard Minerals and Chemicals Corporation in November 1980. It provided an economic comparison for catalytic and thermal incineration for a plastic printing operation and for a formaldehyde plant.

The purchase costs of the catalytic units were based on actual costs for incinerator systems Englehard had previously supplied. These costs included not only the catalytic incinerator, but a heat exchanger, a preheater, a fan, and a high temperature valve as well. Installation costs were assumed to be equal to the purchase costs. Total installed costs were higher for the printing operation than for the formaldehyde plant primarily because of the requirements for a fan and the need for a high temperature valve to partially duct clean incinerator exhaust gas back to the ovens. Annual costs were less for the formaldehyde plant because of the high concentration of VOC in the waste gas stream.

Total installed capital costs were approximately fifty percent higher in Englehard's study than in our study. The purchase cost of the Englehard unit was approximately 15 percent higher than the purchase cost used in our study. The author of Englehard's study noted that this company's systems are generally more expensive than its competitors' and attributed it to the catalysts which contain a greater amount of precious metal. More importantly, the installation cost factor was 67 percent higher than the cost factor used in the cost methodology in this report. By subtracting the cost of the supplemental fuel tanks which was included in their installation costs the factors were approximately the same. The annual costs for the two cases they examined were very close to those calculated by the methodology in this report. Englehard's costs tended to be about 10 percent lower.

The conclusions of Engelhard's study confirmed this report's conclusion that capital costs for catalytic incinerators are typically slightly higher than those for thermal incinerators and that annual costs are generally substantially lower. Capital costs for catalytic systems ranged from 5 to 20 percent higher than those for thermal systems, whereas the annual costs were about 30 percent lower for low concentrations. However, Engelhard's conclusions contradicted this report's findings that for high concentrations, and especially with heat recovery, thermal incineration is more economical. Engelhard concluded that catalytic incineration is the lower cost technology even at high VOC concentrations. This result is believed to be due to the assumptions for supplementary fuel use. Engelhard assumed no supplementary fuel use for catalytic incineration and a substantially higher supplementary fuel use for thermal incineration than is predicted by this report.

Previous Radian Corporation Study

Radian Corporation performed a confidential evaluation of various VOC control technologies for a chemical operation. Cost estimates were developed for the following: fixed-bed carbon adsorption with solvent recovery, fixed-bed carbon adsorption with thermal incineration, fluidized-bed carbon adsorption, thermal incineration at various heat recoveries, and catalytic incineration with and without heat recovery. Capital and annualized costs for catalytic incineration were both about 20 percent higher than those generated in Section 4.3.1. This discrepancy can be attributed to the inherent differences between chemical processes and solvent evaporation processes.

The study concluded that the most economical means for achieving the desired degree of VOC control for this particular process were carbon adsorption with solvent recovery and thermal incineration with regenerative heat exchange. It must be mentioned that this particular plant operated continuously throughout the year thus enhancing the suitability of the regenerative system, and the waste gas contained only one component thus making the carbon adsorption system more economical.

REFERENCES

1. U. S. Environmental Protection Agency. Control of Volatile Organic Emissions from Existing Stationary Sources, Volume I: Control Methods for Surface-Coating Operations. (Prepared for U. S. Environmental Protection Agency). Research Triangle Park, N.C. EPA Publication No. 450/2-76-028. November 1976.

2. U. S. Environmental Protection Agency. Control Techniques for Volatile Organic Emissions from Stationary Sources. Research Triangle Park, N.C. Publication No. EPA-450/2-78-022. May 1978. p. 75.

3. U. S. Environmental Protection Agency. Publication Rotogravure Printing - Background Information for Proposed Standards. Research Triangle Park, N.C. Publication No. EPA-450/3-80-031a. October 1980. Chapter 4.

4. U. S. Environmental Protection Agency. Standards of Performance for New Stationary Sources: Graphic Arts Industry; Publication Rotogravue Printing; Proposed Rule and Notice of Public Hearing. Federal Register, Vol. 45: 71538-71556, Part IV. Washington, D.C. U. S. Governmental Printing Office. October 28, 1980.

5. U. S. Environmental Protection Agency. Pressure Sensitive Tape and Label Surface Coating Operations; Standards of Performance for New Stationary Sources; Public Hearing and Proposed Rulemaking. Federal Register, Vol. 45: 86278-86293, Part VIII. Washington, D.C. U. S. Government Printing Office. December 30, 1980.

6. Letter and attachments from Goodwin, D. R., EPA:ESED, to Waldron, H., General Tire and Rubber Company, and list of addressees. August 10, 1981. 1 p. notice of NAPCTAC meeting and request for review of draft preamble, regulation and background information document.

7. General Motors. Report on Phase I Testing and Phase II Continuance Pursuant to a Delayed Compliance Order Issued on February 15, 1980 by the United States Environmental Protection Agency, Governing the GM Assembly Division at Fremont, California. (Prepared for U. S. Environmental Protection Agency, Region IX.) 1980.

8. U. S. Environmental Protection Agency. Flexible Vinyl Coating and Printing Operations -- Background Information for Proposed Standards; Draft. Research Triangle Park, N.C. EPA Contract No. 68-02-3058. August 1981. Chapter 4.

9. Parmele, C. S., W. L. O'Connell, & H. S. Basdekis. Vapor-phase Adsorption Cuts Pollution, Recovers Solvents. Chemical Engineering. 86(28):59-70. December 31, 1979.

10. Letter and attachment from Ametek, to Burt, R., Rohm & Haas Company. February 10, 1975. 52 p. Information to Ametek's scrubbing equipment and enclosure of Catalog 7R.

11. Rolke, R. W., et al. Afterburner Systems Study. (Prepared for U. S. Environmental Protection Agency.) Research Triangle Park, NC. Publication No. EPAR272062. August 1972. p. 15-17.

12. Reference 1, pp. 38-51.

13. Reference 1, pp. 68-92.

14. Full Scale Carbon Adsorption Study for Radian Corporation for the Industrial Environmental Research Laboratory. Cincinnati, Ohio, EPA Contract No. 68-03-3038.

15. Hobbs, F. D. C. S. Parmele, and D. A. Barton. (IT Enviroscience, Inc.) Survey of Industrial Applications of VaporPhase ActivatedCarbon Adsorption for Control of Pollutant Compounds from Manufacture of Organic Compounds. (Prepared for the U. S. Environmental Protection Agency.) Research Triangle Park, NC. EPA Contract No. 68-03-2568. April 1980.

16. Meinhold, T. F. Fume Incineration for Air Pollution Control. Plant Engineering. 34(23):108-115. November 13, 1980.

17. Letter and attachment from Pennington, R. L., Regenerative Environmental Equipment Co., Inc., to Krohn, N. E., Radian Corporation. October 7, 1981. 4 p. Information on the REECO RETHERM system.

18. OxyCatalyst, Inc. Oxycat Catalytic Abatement Systems; Gas Fired. [1980.]

19. Neveril, R. B. (GARD, Inc.) Capital and Operating Costs of Selected Air Pollution Control Systems. (Prepared for U. S. Environmental Protection Agency.) Research Triangle Park, NC. EPA Publication No. 450/5-80-002. December 1978.

20. Reference 3, Chapter 8.

21. CE Air Preheater. Report of Fuel Requirements. Capital Cost and Operating Expense for Catalytic Afterburners. (Prepared for U. S. Environmental Protection Agency.) Research Triangle Park, NC. EPA Publication No. 450/3-76-031. Septebmer 1976.

22. U.S. Environmental Protection Agency. Organic Chemical Manufacturing, Volume 4: Combustion Control Devices. Research Triangle Park, NC. EPA Publication No. 450/3-80-026. December 1980.

23. Price: Natural Gas, Electricity. Monthly Energy Review. 81:85-86. October 1981.

24. Trip Report. Jennings, M. S. and N. E. Krohn, Radian Corporation, to file. November 28, 1981. 4 p. Report of visit to DuPont plant on November 2, 1981, in Wilmington, Delaware.

25. Memo to file. Robert Berry, Radian Corporation. December 15, 1981. Thermal Incineration and Carbon Adsorption Costing Methodology for Representative Sources.

26. Martin, N. A. (Engelhard/Systems). Catalytic Incineration of Low Concentration Organic Vapors. (Prepared for U. S. Environmental Protection Agency.) EPA 600/2-81-017 (NTIS PB81-158441). Research Triangle Park, North Carolina.

5. Determination of Factors Indicating Suitability of Catalytic Incineration

The previous chapter presented a description of alternative VOC control techniques and examined technical constraints and costs of carbon adsorption, thermal incineration and catalytic incineration. Drawing from this analysis, this chapter presents a generalized guide to applications when catalytic incineration is the most suitable VOC control option.

This chapter approaches the question of suitability from two perspectives. The first perspective reviews the applications where catalytic incineration is now used and provides an indication of suitability as a function of industry and application. The second perspective examines each process and emission characteristic affecting suitability and characterizes the applications where catalytic incineration is a suitable VOC control technique. The combination of these two perspectives identifies the factors indicating suitability of catalytic incineration.

5.1 REVIEW OF PRESENT APPLICATIONS

As discussed in Section 3.3, the majority of catalytic incinerators used for control of VOC emissions are found in the surface coating industries. Of the plant sites using catalytic incinerators identified in this study, 60 percent are surface coating operations. Usually these incinerators control oven or dryer emissions. Of the catalytic incinerators identified in this study in which the process type is available, approximately 30 percent are used to control oven or dryer emissions.

In the surface coating industries, ovens or dryers are generally the most easily controlled VOC emission sources. Other VOC emission sources are also found in the surface coating industries, however, these sources are not as readily controlled. Oven and dryer emissions are easily ducted and directed to control devices. VOC concentrations in the oven or dryer waste gas streams are usually less than 25 percent LEL; 5 to 20 percent LEL is typical. Temperatures commonly range between 200 and 600°F (see Section 4.3). These concentrations and temperatures are very compatible with the operating conditions of catalytic incinerators. In many of these waste gas streams, the heat produced and recovered from the combustion of the VOC's reduces or eliminates the supplementary fuel used by the catalytic incinerator preheat burner. This reduction in preheat fuel can significantly reduce control costs compared to other control alternatives. Also, in some cases, heat recovery through recycle can reduce the combined fuel consumption of the process and incinerator resulting in a net energy savings.

As discussed in Section 3.3, catalytic incinerators are often used to control waste gas streams containing multiple VOC compounds. Catalytic incinerators are well suited to controlling both single and multicomponent VOC waste gas streams. Streams containing multiple VOC compounds are sometimes difficult to control with VOC recovery methods such as carbon adsorption. Because of the complexity of the VOC separation and purification process, use of carbon adsorption on multicomponent VOC streams may be considerably more costly than catalytic incineration depending on the value of recovered compounds.

Catalytic incinerators are generally used to control waste gas streams containing very low concentrations of particulate matter and catalyst poisons. Particles and poisons can erode, mask or deactivate the catalyst and either significantly reduce the lifetime of the catalyst or increase maintenance of the catalyst. Low levels of potential poisons or particulates can be tolerated depending on the types of poisons or particles and the incinerator operating conditions.

5.2 PROCESS AND EMISSION CHARACTERISTICS AFFECTING SUITABILITY

A description of the primary factors affecting suitability of catalytic incineration is presented in Table 5-1. For the purpose of this analysis, factors affecting suitability have been divided into three categories:

- Waste Gas Characteristics
- Performance Requirements
- Other Factors

5.2.1 Waste Gas Characteristics

The waste gas characteristics, namely the waste gas temperature, VOC concentration, and VOC species, have a major effect on the suitability of catalytic incineration. All three factors affect the supplemental fuel requirements of the incinerator which in turn affects control costs. As discussed in Section 3.1, the waste gas usually must be heated up to a minimum temperature to initiate oxidation. The amount of heating required is dependent on the waste gas temperature, VOC concentration and VOC species, the amount of heat recovery used, and the VOC destruction efficiency of the incinerator.

Figure 5-1 shows the variation in supplemental fuel usage for various waste gas temperatures and VOC concentrations. The unshaded area on the graph identifies the region where auxiliary fuel will probably not be needed except for startup. In this area, catalytic incineration is particularly well suited. For VOC concentrations below about 12 percent LEL, supplementary fuel is generally required. For VOC concentrations between about 12 and 23 percent LEL, no preheating is required if: (1) a heat exchanger is used to preheat the incoming waste gas with the hot, incinerator exhaust gas and (2) the waste gas temperature is at least 200°F.

Table 5-1. Summary of the primary factors affecting the suitability of catalytic incinerators.

	Waste Gas Characteristics	Performance Requirements		Other Factors	
Waste Gas Temperature	At low LEL's, waste gas must be heated to 900°F[a] At higher LEL's, waste gas must be heated to 600°F[a]	VOC removal efficiency	Greater than 95+% depending on VOC type and VOC concentration	Space Limitations	Units are small, lightweight and can be elevated
VOC Concentrations	Up to 25 percent LEL	Exhaust gas VOC Concentration	Insufficient data	O&M personnel	0.5 manhours/ 8 hr shift
VOC Compounds	Virtually all; if VOC is relatively high value (greater than 0.75 to 1.00 $/gal) carbon adsorption favored	Energy usage	If WGT <200°F, preheat fuel is required	Secondary emissions	15 ppm thermal NO_x
Contaminants	Less than 25 ppm of potential catalyst poisons[3] Less than 0.05 gr/SCF of particulate matter [3]		If WGT is 200°F-600°F, no preheat is required for streams greater than 10% LEL if heat recovery is used If WGT >600°F, no preheat fuel is required except for low LEL streams	Water pollution Solid Waste	None Negligible

[a]Using supplementary fuel or heat exchange with exhaust gas.

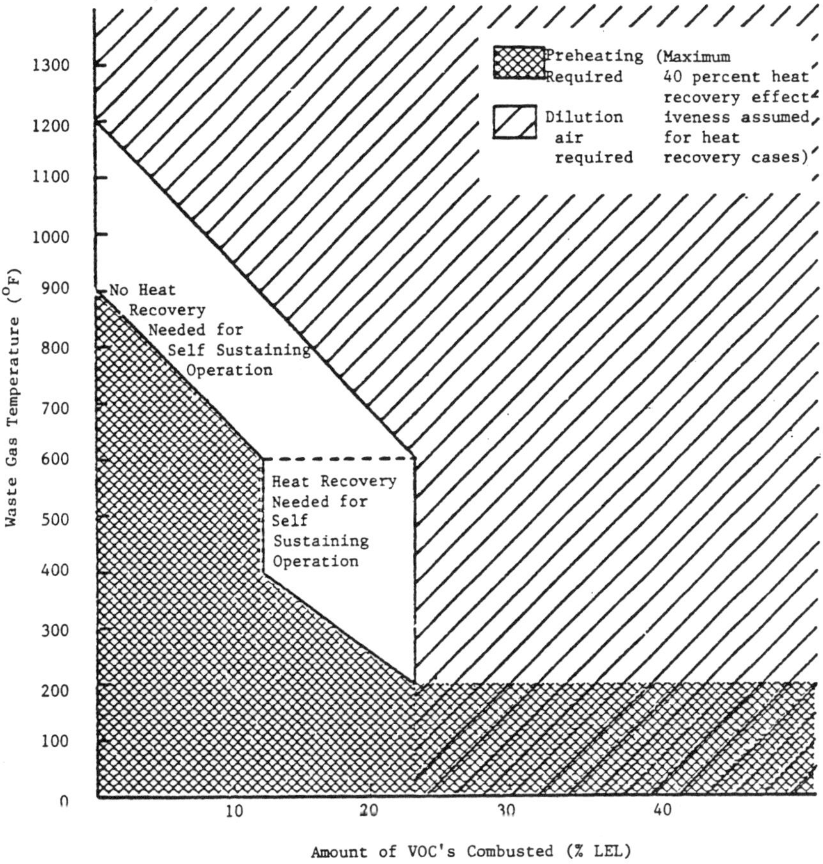

Figure 5-1. Supplementary fuel and dilution air requirements.

VOC concentrations greater than about 23 percent LEL will usually liberate enough heat upon combustion that dilution air, to prevent catalyst overheating, will be required. Dilution air requirements are dependent on a number of factors including the type of catalyst and the VOC species.

Addition of dilution air decreases the VOC concentration and increases the size of the catalytic incineration equipment. Sometimes, adding dilution air can increase costs significantly. Thermal incinerators operate at higher temperatures and thus do not require dilution air to reduce temperatures to acceptable levels. Thus, thermal incinerators may be less costly for applications in which dilution air would be required to use catalytic incineration. Figure 5-2 illustrates two specific examples. The upper two lines compare unitized annualized costs ($\$/10^6$ scf) for catalytic and thermal incineration (no heat recovery) of a waste gas stream at 100°F. The lower two lines present the same comparison for a waste gas stream at 600°F. As the VOC concentration increases, the control costs for thermal incineration decreases almost linearly due to the reduction in fuel use. Because no heat recovery is used, the supplementary fuel requirements of the catalytic incineration cases are constant until dilution air must be added. For these increases, control costs are essentially constant until dilution air is required. At the point where dilution air is required, fuel requirements increase and costs rise sharply. A crossover point (the point where control costs are equal) occurs between 20 and 25 percent LEL. The crossover point is dependent on a number of factors including the waste gas temperature, the amount of heat recovery used in the system, the VOC concentration and VOC species.

As stated previously, catalytic incinerators are well suited to controlling streams with multiple VOC compounds. The high cost of recovery of multiple VOC compounds may preclude the use of carbon adsorption. However, if the recovery credit is high enough, carbon adsorption can be a less costly alternative. Figure 5-3 illustrates the recovery credit sensitivity. The graph shows unitized annualized costs ($\$/10^6$ scf) for carbon adsorption and catalytic incineration as a function of recovery credit. Distillation and decantation recovery methods are shown because in many situations, decantation is the less costly recovery method. As the recovery credit increases, the costs for the recovery system decreases. Catalytic incineration costs are unaffected by recovery credit. For recovery credits above about $0.75/gallon, carbon adsorption using decantation is less costly than catalytic incineration. For recovery credits above about $1.00/gallon, carbon adsorption using distillation is less costly than catalytic incineration.

It should be noted that variability in the VOC species contained in the waste gas may make recovery of even high value solvents uneconomical. Incinerators can accommodate changes in the VOC species contained in the waste gas as process formulation changes or new products are manufactured. Carbon adsorption systems are much less flexible and may require major

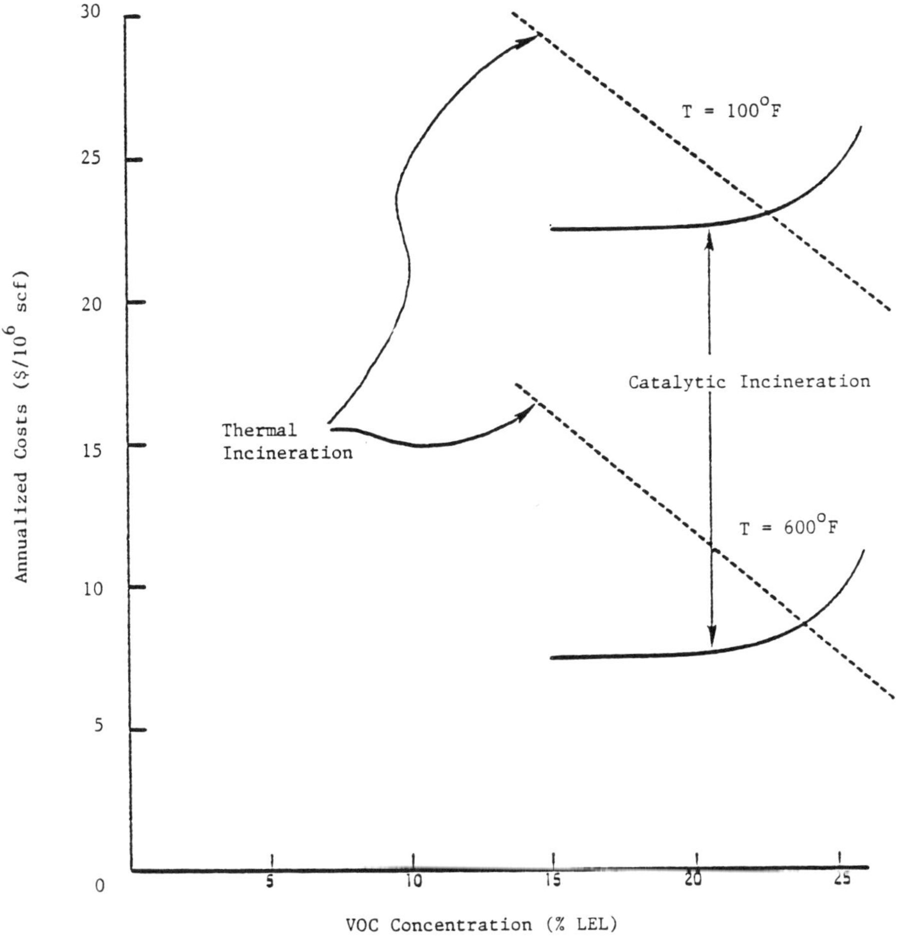

Figure 5-2. Annualized costs of thermal and catalytic incineration for two different waste gas temperatures.

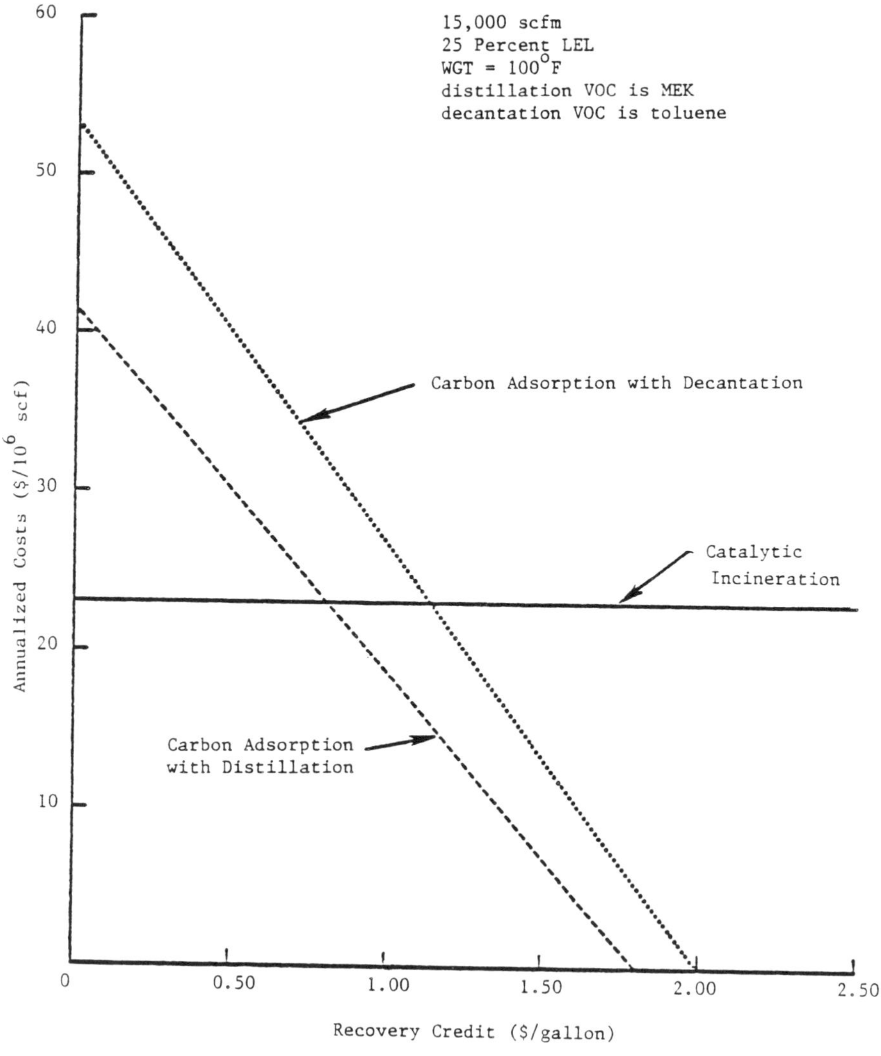

Figure 5-3. Recovery credit sensitivity.

modifications in distillation or decantation recovery equipment when formulations are changed.

5.2.2 Performance Requirements

The performance of a control device can be measured by a variety of criteria. However, for the purpose of this report, the three primary performance criteria are the following:

- VOC destruction efficiency,
- VOC exhaust gas concentration, and
- Energy usage.

This section examines each in turn.

Usually, a control device is required to meet the performance standards of the governing state or local environmental agency. These performance standards vary but most often the control device must achieve a minimum VOC destruction efficiency and/or a maximum VOC exhaust gas concentration. Several emission test reports on catalytic incinerators used for VOC control were obtained from plants and various environmental agencies (see Section 3.1.4). These test reports showed VOC destruction efficiencies ranging from 42.5 to 99 percent, averaging 95.6 percent. Generally the highest efficiencies (99 percent or greater) are only achievable when the waste gas is relatively rich in VOC's. Emission data on VOC exhaust gas concentrations ranged from 15 to 818 ppm. No information concerning the lowest VOC exhaust gas concentration achievable with catalytic incinerators was located during this study.

The major part of the energy used by a catalytic incinerator is in the form of supplemental fuel to the preheat burner. As discussed previously, the supplemental fuel requirements will vary depending on the waste gas temperature, VOC concentration, the VOC species, and the amount of heat recovery used in the system. The supplemental fuel required by a catalytic incinerator is usually much less than that required by a thermal incinerator. The analysis presented in Section 4.3 indicates that catalytic incinerators require at least 25 percent less energy than thermal incinerators in most applications. In some cases, a catalytic unit will operate without supplemental fuel and/or reduce process fuel consumption. The actual energy usage will vary depending on the specific application and the incinerator design (see Section 4.3 and Appendix A).

5.2.3 Other Factors

A variety of other factors may affect the suitability of catalytic incineration. Size is an important consideration in retrofit cases, when a control device is to be installed on an existing piece of equipment or process. Catalytic incinerators are generally lighter weight and smaller than either thermal incinerators or carbon adsorption systems and can be used for roof installations. A typical catalytic incinerator designed for 10,000 SCFM of waste gas and similar in design to Figure 3-1, would weigh between 12,000 and 14,000 lbs. A comparable thermal incinerator would weigh between 20,000 and 27,000 lbs.[3,4]

Operation and maintenance requirements can also be an important suitability consideration. O&M requirements for catalytic, as well as thermal incineration are low; approximately 0.5 manhours per shift. Carbon adsorbers require higher O&M, approximately 1 manhour per shift. Nearly all plants could support the O&M requirements of a catalytic incinerator.

Secondary emissions, in the form of thermal NO_x, are low; about 15 ppm is typical for catalytic incinerators.[5] No water pollution is generated by the operation of a catalytic incinerator. Small amounts of solid wastes are generated by catalytic incinerators in the form of old catalyst support. These low secondary and solid waste emissions should have a negligible effect on the suitability of catalytic incineration in nearly all applications.

REFERENCES

1. Telecon. Martin, Norman, Engelhard with Krohn, N. E. Radian Corporation 1/7/82. Conversation about catalyst's tolerances.

2. Telecon. Madden, Gerald, E.I. DuPont de Nemours, with Krohn, N. E., Radian Corporation. January 7, 1982. Conversation about catalyst tolerances.

3. Telecon. Madden, Gerald, E.I. DuPont de Nemours, with Krohn, N.E., Radian Corporation. January 21, 1982. Conversation about size and weight of incinerators.

4. Rolke, R. W. (Shell Development Company). Afterburner System Study. Report done for the U. S. EPA, Research Triangle Park, NC). EPA-R2-72-062. August 1972. p. 79.

5. Reference 4, p. 7.

6. Measurement Methodology for Field Testing Catalytic Incinerators

This chapter provides general methodologies for field testing catalytic incinerators. The methodologies presented provide a scheme for assessing the overall performance of catalytic incinerators in various applications. As described in Chapter 3, the performance of incinerators can be measured in a number of ways, including VOC destruction efficiency, VOC outlet concentrations, and energy usage.

In order to adequately assess overall performance, it is necessary to characterize the VOC emitting process as well as the actual incinerator system. Thus, the methodologies must include provisions to characterize and monitor the process and the incinerator during the emission test. In general, this requirement will be highly site specific in nature. This chapter only highlights major elements of this aspect of the emission test.

The first two subsections of this chapter discuss the characterization of the VOC emitting process and the incinerator, respectively. Following this, the available VOC emission testing methodologies are reviewed. Finally, the chapter concludes with a brief review of other testing considerations.

6.1 PROCESS CHARACTERIZATION AND MONITORING

A characterization of the VOC emitting process is required in order to assess the performance of a catalytic incinerator in relationship to process emissions. Characterization is required for the following reasons:

1. To determine if the process is operating within normal day-to- day tolerances during the emission test. Generally, periods in which a process upset is occurring are avoided during the test.

2. To characterize the degree and magnitude of fluctuations in flow rates, temperatures, pressures, etc. These will be particularly important for processes which inherently fluctuate (i.e., coating processes) since performance during non-steady state operation can be analyzed.

3. To provide a process data base useful in predicting catalytic incinerator performance on other similar emission sources.

4. To provide background data useful in assessing anomalies in emission test results.

It is important to note that the required process data and monitoring scheme will be highly site specific. In addition, industrial plants may consider process data confidential and limit the type and amount of data which can be collected. Both of these factors must be considered during test site selection. Furthermore, a reasonable time period must be planned to allow adequate pretest site visits and to familiarize personnel with the overall production process.

Table 6-1 presents a list of information needed for general process characterization. These data will typically be gathered during a pretest site survey. Information concerning the type of VOC emitting process, the process configuration and the number of process sources feeding the incinerator must be known for sampling site determination. Sampling sites for measuring volumetric gas flows and VOC concentrations in the stacks and ducts are usually selected using strict criteria contained in EPA Reference Method 1.[1] Sometimes stack extensions or alterations are required to meet these criteria. Such alterations are usually costly; therefore, sampling site determination is a key task in the pretest survey.

Information concerning the process conditions, process control system and monitoring system is needed in order that testing instrumentation requirements may be assessed. In many instances, all or part of the required testing instrumentation, such as temperature and pressure monitors, are already present in the process. Use of existing process monitors and controls, if determined to be operating properly, will reduce the number of test instruments the testing team will need to supply.

In order to prepare a site specific test plan, the length of the testing period must be set. Knowledge of the design production rates and production schedule is necessary to determine the test length. For example, if a process line is up for 2 hours and down for 30 minutes as a normal operating schedule, the testing period must be at least 3 hours to include all fluctuations.

The actual process monitoring scheme used during the test program will be highly site specific. Table 6-2 is presented as a general guide to the information which should be gathered. The size and complexity of the process are major factors influencing the degree of process monitoring conducted. For simple processes, monitoring of a few key temperatures may suffice. For more complex processes the monitoring effort will be more extensive. Operating personnel at the plant site are usually the best guides in the determining appropriate monitoring requirements.

In general, the process should be monitored for a period before the emission test and then during the test. For continuous emission tests conducted over several days or weeks, process data should be gathered continuously with the test length depending on the fluctuations in the process.

TABLE 6-1. PROCESS INFORMATION NEEDS

1.	Type of VOC emitting process
2.	Process configuration
3.	Number of sources feeding incinerator selected for testing
4.	Design temperatures, flowrates, temperatures, etc. (and expected fluctuations in each)
5.	Control system configuration
6.	Monitoring system configuration
7.	Design production rates
8.	Production schedule

TABLE 6-2. PROCESS MONITORING REQUIREMENTS

1.	Process temperatures
2.	Process pressures and pressure drops
3.	Solvent(s) flowrates
4.	Solvent(s) levels
5.	Production rate
6.	Process fuel use (i.e. oven gas consumption)

6.2 INCINERATOR CHARACTERIZATION AND MONITORING

Characterization of the incinerator is another element in the data gathering effort required before and during an emission test. The major emphasis in characterizing and monitoring the incinerator is to relate the overall incinerator performance to changes in key variables such as incinerator inlet and exit temperatures and VOC concentrations. Catalytic incinerators can typically achieve a very wide range of VOC destruction efficiencies and outlet VOC concentrations depending on a number of factors. Thus, the best measure of performance is the VOC destruction efficiency and exhaust gas VOC concentration achieved under a given set of operating conditions.

Table 6-3 details design data and incinerator information which would typically be required prior to an emission test. Information concerning the catalyst and date of installation is useful in assessing the effects of catalyst life and equipment life on overall performance. Incinerator instrumentation requirements during the test can be assessed with information concerning the controls, monitors and safety features of the catalytic incinerator.

Table 6-4 details important incinerator operating conditions which must be monitored during the emission test. The waste gas temperature, catalyst bed temperature, and exhaust gas temperature are excellent indicators of normal incinerator operation and heat recovery effectiveness. For example, measurement of the gas temperature rise across the catalyst bed gives an indication of the VOC destruction efficiency (i.e., for typical solvents 26.5°F for each percent LEL combusted). Monitoring of the pressure drop across each incinerator element (heat exchanger, bed, etc.) allows a pressure profile of the catalytic incinerator system to be seen. Fuel consumption monitoring allows a heat balance to be performed around the incinerator.

Figure 6-1 shows the placement of the temperature and pressure monitors around the catalyst bed in a typical catalytic incineration system. Catalytic incinerators using a type of heat recovery other than the recuperative heat exchanger shown in Figure 6-1 (e.g. recycle or waste heat boiler) would require different monitor placement. Generally, the site specific test plans will detail the placement of monitors to obtain the necessary information to perform a heat balance on the total system.

6.3 EMISSION MEASUREMENT PROCEDURES

The VOC destruction efficiency of a catalytic incinerator is determined from the relative difference between the mass of VOC in the waste gas stream and the mass of VOC in the exhaust gas stream exiting the catalyst bed. This relation is expressed in the following equation.

TABLE 6-3. CATALYTIC INCINERATOR INFORMATION NEEDS

1. Manufacturer and model of incinerator
2. Date of installation
3. Manufacturer and type of catalyst
4. Catalyst age, volume, and cleaning/regeneration history
5. Incinerator configuration and heat recovery scheme
6. Design operating temperatures, pressures, flowrates, etc.
7. Control and monitoring system
8. Safety features

TABLE 6-4. CATALYTIC INCINERATOR PROCESS MONITORING REQUIREMENTS

1. Waste gas temperature
2. Gas temperature prior to catalyst bed
3. Gas temperature after catalyst bed
4. Exhaust gas temperature
5. Pressure drop across each incinerator section (preheat, bed, heat exchanger)
6. Fuel consumption/composition

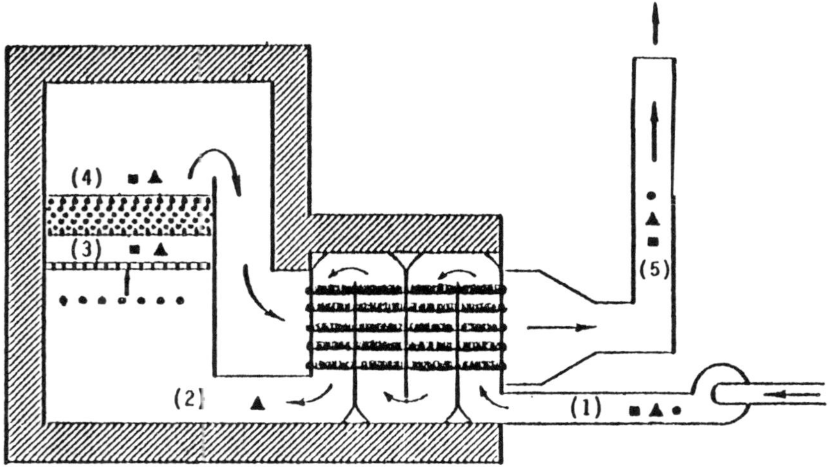

■ temperature monitor
▲ pressure monitor
● sampling point for measurement of VOC concentration and gas flow rate

1. Waste gas temperature, pressure, VOC concentration and flow rate
2. Waste gas pressure after heat exchanger
3. Gas temperature and pressure prior to the catalyst
4. Gas temperature and pressure after the catalyst
5. Exhaust gas temperature, pressure, VOC concentration and flow rate

Figure 6-1. Monitor placement.

$$\text{Efficiency (\%)} = \frac{VOC_{in} - VOC_{out}}{VOC_{in}} \times 100 \qquad \text{Eq. (6-1)}$$

Where VOC_{in} = mass flowrate of VOC contained in the waste gas
VOC_{out} = mass flowrate of VOC contained in the exhaust gas

The mass of VOC contained in the gas streams is found by measuring the volumetric flow rate and VOC concentration of the gas stream for an appropriate time period in both the inlet and outlet of the incinerator. In cases where preheat fuel and/or dilution air is used, the inlet and outlet flow rates will differ. Direct measurement of the exhaust gas VOC concentration provides another measure of catalytic incinerator performance.

To insure that all emission measurements are collected under well defined and commonly used procedures, EPA reference methods are most appropriate. A combination of EPA Reference Methods 1-4 can be used to measure the volumetric flow rate. EPA Reference Method 1, "Sampling and Velocity Traverses for Stationary Sources", is used to select the correct sampling site for measuring the volumetric flow rate in the ducts or stacks.[1] EPA Reference Method 2, "Determination of Stack Gas Velocity and Volumetric Flow Rate", is used for measuring the gas flow rate.[2] In order to adjust the gas flow rate to dry standard conditions, the molecular weight and moisture content of the gas must be known. EPA Reference Method 3, "Gas Analysis for Carbon Dioxide, Oxygen, and Excess Air", is used to determine the molecular weight.[3] EPA Reference Method 4, "Determination of Moisture Content in Stack Gases", is used to measure the moisture content.[4]

Four EPA reference methods are applicable for the measurement of VOC concentrations:

Method 18 (draft), "Measurement of Gaseous Organic Compound Emissions by Gas Chromatography."[5]

Method 25, "Determination of Total Gaseous Nonmethane Organic Emissions as Carbon."[6]

Method 25A, "Determination of Total Gaseous Organic Concentration Using a Flame Ionization Analyzer."[7]

Method 25B, "Determination of Total Gaseous Organic Concentration Using a Nondispersive Infrared Analyzer."[8]

EPA Reference Method 25 involves the collection of an integrated grab sample. The time period of collection may be varied to any appropriate time interval; however, most tests involve sampling periods on the order of an hour. Because each sample requires an extensive laboratory analysis, the collection of numerous grab samples over shorter time intervals can be very expensive. Recently, on-site portable Method 25 analyzers have become available; however, operating experience is limited. These portable analyzers collect a sample over several minutes. Regardless of the time interval, a single VOC concentration is measured which is an average over the sampling period.[9] Because of this, continuous emission measurement methods (Method 25A) are more appropriate for fluctuating processes where a continuous measurement of VOC concentration is desired. Problems have also been encountered with Reference Method 25 when measuring VOC concentrations of streams with high CO_2 concentration.

EPA Reference Method 25B is not typically suitable for VOC measurement of streams containing high concentrations of water vapor or carbon dioxide.[10] Since both water vapor and carbon dioxide are combustion products present in catalytic incinerator exhaust streams, EPA Reference Method 25B is not a suitable VOC measurement method for catalytic incineration testing.

Both EPA Reference Method 25A and draft Method 18 are suitable methods for measuring VOC concentrations in gas streams associated with catalytic incinerators. Method 18 (draft) can identify and quantify individual VOC species and concentrations through gas chromatograph analysis.[5] Compound specific information provided by Method 18 is useful in determining (1) if partial oxidation of the VOC compounds occurs and (2) for multicomponent VOC streams, if selective destruction of certain VOC's occurs.

EPA Reference Method 25A can provide continuous on-line measurement of the VOC concentration, allowing process fluctuations to be related to the performance of the incinerator. This is important because catalytic incinerators are often used to control processes such as coating lines which have fluctuating VOC emissions.

Method 25A measures total hydrocarbon concentration using a flame ionization analyzer (FIA). The principle of the FIA is based on the linear relationship of ion formation to the concentration of a given organic compound. Inside the analyzer, the test sample is burned in a small jet with air or oxygen and the change in the electrical conductivity produced by the combustion of the sample is measured. The output of the FIA is related to VOC concentration using a calibration curve.[11]

FIA's are normally calibrated using single component standards such as propane in air or nitrogen. Figure 6-2 is an example of a typical calibration curve. An FIA output of 0.5 would be equivalent to 100 ppm VOC (as propane). However, in order to determine the actual VOC concentration, the composition of the stream must be known. This is because the FIA's response (i.e. relative to propane) to each particular compound depends on the molecular structure of the compound. Oxygenated compounds such as ketones and aromatic hydrocarbons such as toluene respond differently. The results provided by draft Method 18 can be used to convert the FIA output to the actual VOC concentration. The combined results of Methods 18 and 25A can provide accurate, continuous measurement of VOC mass emission rates.[11]

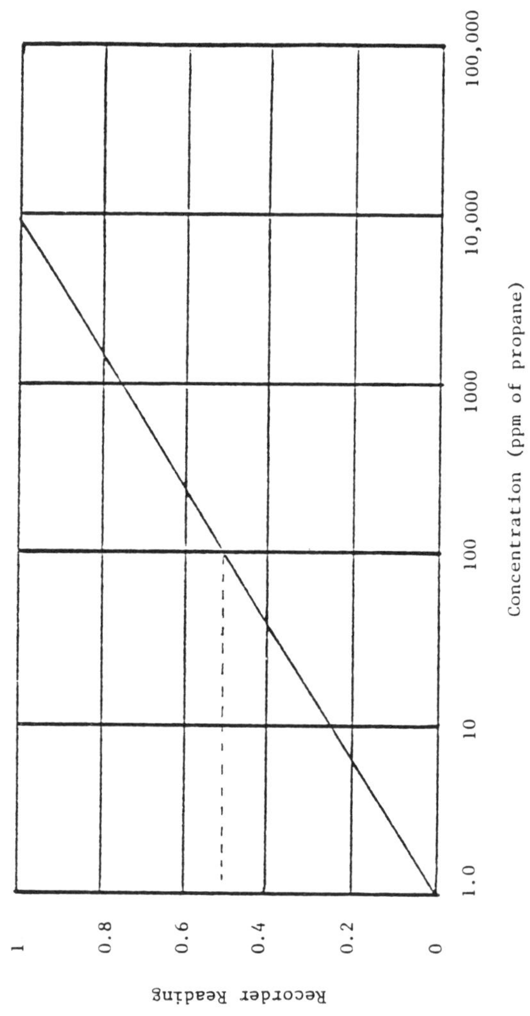

Figure 6-2. Example of VOC calibration curve.

6.4 OTHER CONSIDERATIONS

Test Periods

A majority of the processes using catalytic incineration for VOC control are coating lines. Many coating lines operate intermittently (e.g. up for 10 minutes down for 15 minutes, up for 10 minutes, down for 10, etc.). These short up and down periods are usually required to adjust the process to meet the product specifications. Continuous monitoring of VOC concentration will allow the effect of the intermittent operation to be seen. The length of time needed for continuous monitoring must be decided before testing begins. The test length or period is decided based on a consideration of costs versus test needs or requirements.

Secondary Emissions

Secondary air emissions are created due to the combustion of sulfur, nitrogen or halogen containing compounds in the catalytic incinerator. If the waste gas directed to a catalytic incinerator contains compounds such as these, secondary air emissions may result. An additional consideration of a test program for catalytic incinerators might be measurement of secondary air emissions. There are several EPA reference methods applicable to testing for secondary emissions:

Method 6, "Determination of Sulfur Dioxide Emissions from Stationary Sources."[12]

Method 7, "Determination of Nitrogen Oxide Emissions from Stationary Sources."[13]

Method 8, "Determination of Sulfuric Acid Mist and Sulfur Dioxide Emissions from Stationary Sources."[14]

None of these methods allow for continuous monitoring and all require lab analysis. Several continuous monitors for NO_x and SO_2 emissions measurement could be used as alternatives to the reference methods and should be considered for this testing program if the need exists. Development of secondary air emissions measurement procedures must be done on a case by case basis.

REFERENCES

1. U. S. Environmental Protection Agency, Code of Federal Regulations. Reference Method 1, Sampling and Velocity Traverses for Stationary Sources. Title 40, Chapter 1, Part 60, Appendix A. Washington, D.C. Office of the Federal Register. July 1, 1981.

2. U. S. Environmental Protection Agency. Code of Federal Regulations. Reference Method 2, Determination of Stack Gas Velocity and Volumetric Flow Rate. Title 40, Chapter 1, Part 60, Appendix A. Washington, D.C. Office of the Federal Register. July 1, 1981.

3. U. S. Environmental Protection Agency. Code of Federal Regulations. Reference Method 3, Gas Analysis for Carbon Dioxide, Oxygen, Excess Air and Dry Molecular Weight. Title 40, Chapter 1, Part 60, Appendix A. Washington, D.C. Office of the Federal Register. July 1, 1981.

4. U. S. Environmental Protection Agency. Code of Federal Regulations. Reference Method 4, Determination of Moisture Content in Stack Gases, Title 40, Chapter 1, Part 60, Appendix A. Washington, D.C. Office of the Federal Register. July 1, 1981.

5. U. S. Environmental Protection Agency. Draft Reference Method 18, Measurement of Gaseous Organic Compound Emissions by Gas Chromatography. Emission Standards and Engineering Division, Research Triangle Park, NC. 1981

6. U. S. Environmental Protection Agency. Code of Federal Regulations. Reference Method 25, Determination of Total Gaseous Nonmethane Organic Emissions as Carbon. Title 40, Chapter 1, Part 60, Appendix A. Washington, D.C. Office of the Federal Register. July 1, 1981.

7. U. S. Environmental Protection Agency. Method 25A, Determination of Total Gaseous Organic Concentration Using a Flame Ionization Analyzer. 45 Federal Register 83149. Washington, D.C. U. S. Government Printing Office. December 17, 1980.

8. U. S. Environmental Protection Agency. Method 25B, Determination of Total Gaseous Organic Concentration Using a Nondispersive Infrared Analyzer. 45 Federal Register 83152-83153. Washington, D. C. U. S. Government Printing Office. December 17, 1980.

9. U. S. Environmental Protection Agency. Measurement of Volatile Organic Compounds. Research Triangle Park, NC. Publication No. EPA-450/2-78-041. September 1979.

10. PedCo Environmental, Inc. Summary of Available Portable VOC Detection Instruments. (Prepared for U. S. Environmental Protection Agency). Washington, D.C. EPA Contract No. 68-01-4147. March 1980.

11. Haaek, L. P. et. al. Gas-Chromatographic Technique for the On-Line Evaluation of Solvent Emission Abatement Devices. Environmental Science and Technology. 15:1463-1467. December 1981.

12. U. S. Environmental Protection Agency. Code of Federal Regulations. Reference Method 6, Determination of Sulfur Dioxide Emissions from Stationary Sources. Title 40, Chapter 2, Part 60, Appendix A. Washington, D. C. Office of the Federal Register. July 1, 1981.

13. U. S. Environmental Protection Agency. Code of Federal Regulations. Reference Method 7, Determination of Nitrogen Oxide Emissions from Stationary Sources. Title 40, Chapter 1, Part 60, Appendix A. Washington, D.C. Office of the Federal Register. July 1, 1981.

14. U. S. Environmental Protection Agency. Code of Federal Regulations. Reference Method 8, Determination of Sulfuric Acid Mist and Sulfur Dioxide Emissions from Stationary Sources. Title 40, Chapter 1, Part 60, Appendix 80. Washington, D.C. Office of the Federal Register. July 1, 1981.

7. Testing Needs

[Note: Subsequent to the writing of this report, a testing program was conducted that corrected many of the deficiencies discussed in this chapter.]

The following chapter examines deficiencies in the present catalytic incinerator performance data base and suggests an approach for a testing program. Along with the measurement methodologies presented in Chapter 6, this information is intended to provide a foundation for the development of a comprehensive program to evaluate the overall performance of catalytic incinerators. This performance evaluation would include both emission reduction capabilities and operational characteristics, particularly total incinerator fuel usage.

7.1 DEFICIENCIES IN THE CURRENT DATA BASE

The deficiencies in the current data base stem primarily from the fact that the data were collected with widely varying objectives. As a result, the prior data base is lacking in several respects including:

- insufficient documentation of results,
- insufficient length of testing periods, and
- widely varying emission test procedures.

The first deficiency noted above results in data of limited value. In many cases only the VOC control efficiency is reported. As discussed in Chapter 3, destruction efficiency is only one of several important measures by which catalytic incineration should be evaluated. Outlet VOC concentration and fuel usage are also relevant measures of catalytic incinerator performance. Other data which is typically <u>not</u> reported in present test results include inlet VOC concentration, waste gas temperatures, incinerator temperatures, space velocities, and heat recovery effectiveness. This lack of documentation tends to limit the usefulness of the data in predicting performance for other emission sources which differ from the tested source.

Secondly, the data have often been acquired over relatively short testing periods. This makes the data susceptible to errors due to process fluctuations. Furthermore, data collected over short testing periods seldom is representative of long-term performance. In most cases, emissions data have been generated through grab samples. Only in a limited number of emission tests has the testing been conducted continuously over an eight hour period or longer.

Finally, the testing procedures used to obtain these data have varied considerably, with some being very dissimilar to EPA reference methods. The methods that have been used to measure VOC concentration include flame ionization, gas chromatography, carbon adsorption, and infrared spectroscopy. The methods for measuring other process parameters have varied widely as well. In addition, the testing procedures are often insufficiently documented. To insure the comparability of data the testing methods should be consistent.

7.2 TEST PROGRAM OBJECTIVES

A catalytic incinerator testing program would help to correct the deficiencies described in the previous subsection. In addition to measuring emission reduction capabilities, the test program would measure a wide variety of operational characteristics of the tested incinerators. This data base of operational information would characterize the emission reduction performance as a function of many variables including fuel usage, incinerator temperature, and inlet VOC concentration.

A test program would measure these criteria accurately and determine the effects that various process parameters have on overall performance. To do this, a test program would comply with the following guidelines: (1) It should document extensively all important process and system data. The data which would be collected is detailed in Chapter 6. (2) All tests would be conducted over suitably long time intervals to insure that the data accurately represent typical plant operations. In most cases, data would be collected over several days or weeks of plant operation. (3) EPA testing methods would be employed to insure accuracy and comparability of information within the data base that is ultimately generated. Chapter 6 provides information concerning the EPA reference methods suitable for measuring VOC concentrations in incinerator inlets and exhausts.

7.3 SITE SELECTION APPROACH

A comprehensive testing program would examine catalytic incinerator performance from two perspectives: (1) it would relate performance to specific industries to which catalytic incineration is being applied, and (2) it would relate performance more generally to a variety of process variables.

In order to relate the performance of catalytic incinerators to type of industry, the testing program should include sites that would provide an adequate cross-section of major industries currently using catalytic incinerators. Table 7-1 summarizes the industries and number of plants within each industry identified as using catalytic incinerators. Approximately 150 plants have been identified. A more detailed list including company names, location, and process parameters can be found in Appendix B.

TABLE 7-1. SUMMARY OF SITES IDENTIFIED

Industry	Number of Sites Identified
Metal can coating	25
Magnet wire enameling	24
Organic chemical manufacturing	20
Printing operations	17
Paper coating	5
Asphalt blowing	5
Alkyd varnish production	4
Tire manufacturing	4
Foundry core production	4
Fabric coating	3
Metal coil coating	3
Other metal coating operations	6
Other non-metal coating operations	8
Miscellaneous	21

It is important to note that this list of plants is a random compilation of sites identified in this study. As such, the relative number of plants identified in any industrial sector may be biased due to the availability of information in the particular sector. Nevertheless, the information suggests that metal can coating, magnet wire production, printing operations, organic chemical manufacturing, and paper coating would be the primary industries considered for selecting sites.

In addition to testing a variety of industries, a testing program would also evaluate the performance in relationship to process variables. Of course, this perspective and the industry specific perspective are interrelated, since the process variables examined depend on the industry selected. However, processes should be selected to span the range of variables that have the most effect on catalytic incinerator performance.

There are many process variables affecting performance, and they can be organized into two categories: waste gas parameters and catalytic incineration parameters. These are presented in Table 7-2.

TABLE 7-2. FACTORS AFFECTING CATALYTIC INCINERATION PERFORMANCE

Waste gas parameters
VOC concentration
temperature
flow rate
particulate matter concentration and size distribution
VOC species
Catalytic incinerator parameters
catalyst volume, age, and type
heat recovery configuration
operating temperature
age of incinerator
annual operating hours and consistency of operation
incinerator design

Organizing a test program to analyze and control each of these factors would be unreasonable. Of these factors, some affect performance considerably more than others. These variables, which shall be labeled primary variables, will be the ones given the greatest consideration in planning a test program. They are:

- heat recovery,
- waste gas VOC concentration,
- waste gas temperature, and
- incinerator operation temperature.

It should be noted that each of these primary variables affects incinerator performance in varying degrees.

The testing sites selected, when taken together, should span the range of values observed for each of these primary variables. This means, for example, that sites should be selected to include processes whose waste gas heat contents range from approximately 2 percent to 25 percent LEL. In this way, values that would be expected to occur in existing and potential applications would be examined.

The other factors listed in Table 7-2 should also be examined as extensively as possible. However, these "secondary" variables should be considered only after the requirements for analyzing the primary variables have been satisfied.

Another important consideration on selection test sites is geographical density. Potentially, all of the test sites could be chosen in one or two areas of high catalytic incinerator plant concentration and still comply with the test program objectives described in Section 7.2. If this were done, the testing costs would be reduced. Figure 7-1 illustrates these areas of high concentration. They include Texas, California, Illinois, and Indiana.

7.4 PLANTS WITH EMISSION TEST DATA

Chapter 3 presents the currently available test information. Table 7-3 summarizes the information that will be helpful in selecting actual test sites. The test sources are organized according to industry. The ranges observed for the four primary variables are also included. Appendix B includes more information about plant sites including information for sites in which no emission data is available.

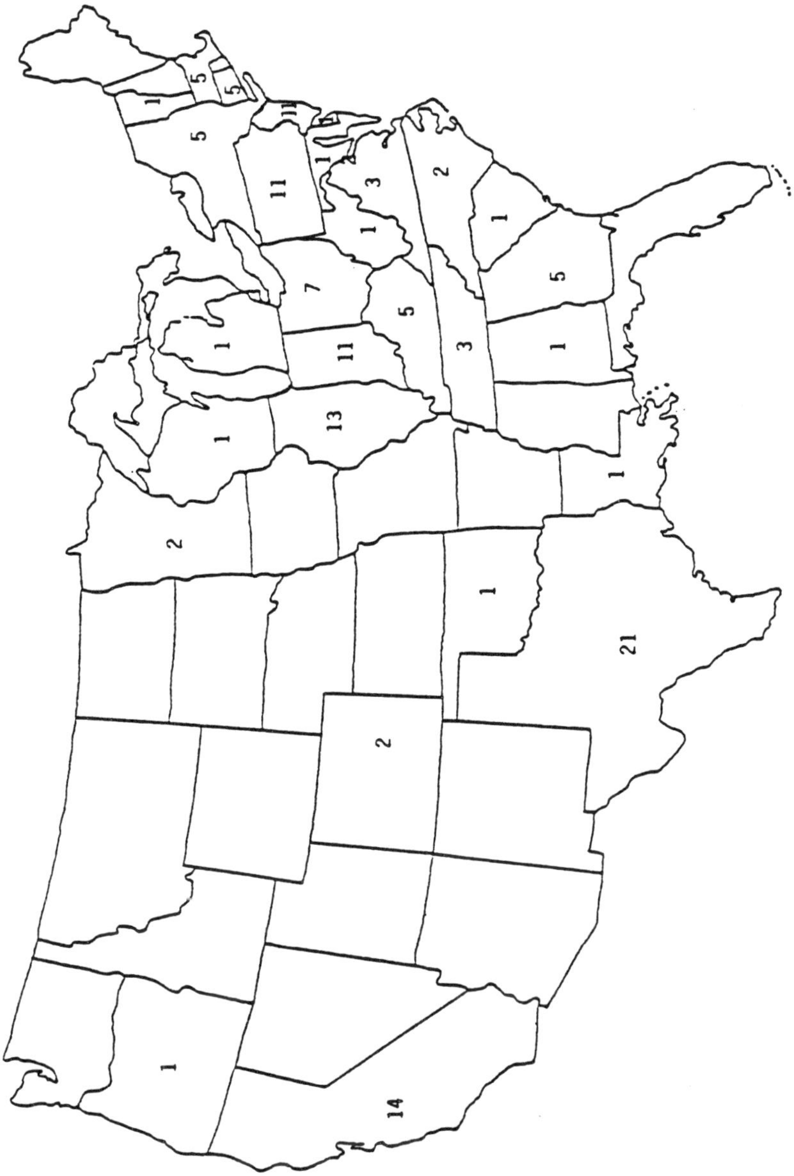

Figure 7-1. Location of plants identified by this study using catalytic incinerators.

TABLE 7.3 SUMMARY OF SITES WITH EMISSION TEST DATA

Industry	Company	Location	Waste Gas VOC Concentration	Primary Variables Heat Recovery	Waste Gas Temp (°F)	Operating Temp (°F)	Other Notes
Metal Can Coating	American Can	Chicago, IL	1300-3000 ppm as CH_4	--	--	730	4 catalytic incinerators
	Continental Can	Los Angeles, CA	5800-9400 ppm as CH_4	yes	290-380	800-445	Heat recovery used
	National Can	Etiwanda, CA	2100 ppm as CH_4	yes	--	--	8 catalytic incinerators
	Western Metal	El Segundo, CA	9000-14000 ppm as CH_4	--	1020	980-1050	
Magnet Wire Production	General Electric	Schenectady, NY	25% LEL	--	--	500-900	Some catalyst beds are internal to the ovens.
		Rome, GA	--	--	--	--	9 incinerators - all external to the ovens
	General Motors Packard Electric Division	Youngstown, OH	--	--	--	--	8 incinerators, 4 have internal beds
Printing Operations	Ambassador College	Los Angeles, CA	4000 ppm as CH_4	no	--	1170	Met-Pro catalytic incinerator
	General Offset	Springfield, MA	--	--	--	--	Torvex catalytic incinerator
	Medallion Graphics	Los Angeles, CA	700 ppm as CH_4	no	170	850	Matthey Bishop catalytic incinerator
	Pacific Press	Vernon, CA	700 ppm as CH_4	--	280	810	

TABLE 7.3 SUMMARY OF SITES WITH EMISSION TEST DATA (Continued)

Industry	Company	Location	Waste Gas VOC Concentration	Primary Variables Heat Recovery	Waste Gas Temp (°F)	Operating Temp (°F)	Other Notes
Petrochemical Manufacturing	DuPont	Beaumont, TX	1.25 wt %	yes	100	750	
	Union Carbide	S. Charleston, WV	--	--	--	--	
Paper Coating	General Formulations	Sparta, MA	1000 ppm	--	--	--	Torvex catalytic incinerator
Tire Manufacturing	B.F. Goodrich	Thomaston, GA	--	--	--	--	Torvex catalytic incinerator
	Uniroyal	Mishawaka, IN	2800-13300 ppm as CH_4	yes	95	800-1000	4 Met-Pro catalytic incinerators
Fabric Coating	Wolverine Fabrics	Blacksburg, VA	--	no	500-600	900	
Coil Coating	Kaiser Aluminum	Toledo, OH	600-4500 ppm	no	--	520	1 catalytic incinerator

Appendix A

Heat and Material Balances Used in Cost Calculations

Appendix A explains the heat and material balances necessary to calculate catalytic incinerator equipment costs and utility costs. This appendix supports the cost methodology detailed in Chapter 4. To determine costs, the incinerator exhaust flow rate must be known. The method for determining exhaust flow rate is presented in Section A.1. With values for exhaust flow rate, natural gas and electricity requirements can be calculated. The methods for determining these utility requirements are presented in Section A.2.

A.1 CALCULATION OF EXHAUST FLOW RATES

A.1.1 Exhaust Flow Rate vs. Waste Gas Flow Rate

The exhaust is defined as the outlet stream from the incinerator. The waste gas is defined as the VOC containing stream, typically from a process oven, which enters the incinerator. The exhaust flow rate is higher than the waste gas flow rate for several reasons.

- Natural gas may be added to the waste gas to preheat the waste gas.
- There are more moles of product than there are moles of reactant.
- At high waste gas temperatures and/or high heat contents, dilution air must be added so as not to exceed the temperature constraints of the catalyst.

A.1.2 The Necessity for Calculating Exhaust Flow Rate

Before sizing a catalytic incinerator, the exhaust flow rate must be known. The exhaust flow rate can be predicted through material and energy balance calculations knowing the percent heat recovery and knowing the following waste gas parameters:

- flow rate
- temperature
- heat content[a]

[a]Heat content depends on both VOC concentration and VOC species. Table A-1 illustrates this relationship for typical solvents.

TABLE A-1. HEAT CONTENTS OF VARIOUS SOLVENTS

VOC Species	Btu/scf/% LEL
Toluene	.56
MEK	.47
Hexane	.61
Benzene	.52

A.1.3 Temperature Constraints

Before these material and energy balances can be performed, constraints concerning maximum allowable operating temperature and minimum combustion temperatures must be established. These constraints must be established to insure reasonable VOC oxidation without catalyst damage and to insure that the VOC begins oxidation upon entering the catalyst bed. These constraints are listed in Table A-2. Figure A-1 illustrates the points within the incinerator system at which these temperature constraints apply by presenting flow schemes for systems both with heat recovery and without heat recovery. These temperature constraints represent typical temperatures for the range of catalytic incinerator applications. The optimum temperatures for a given application depend on a variety of site specific factors, principally the VOC concentration & VOC species.

A.1.4 Graphical Presentation of Exhaust Flow Rate Calculations

By adhering to these constraints and by specifying the percent heat recovery and waste gas parameters, it is possible to calculate the flow rate of the incinerator exhaust. The results of these calculations are presented in Figures A-2 and A-3. The cases are for no heat recovery and 40 percent heat recovery at 95 percent VOC destruction efficiency. The exhaust flow to waste gas flow ratio is plotted against waste gas temperature for various VOC concentrations. Toluene is assumed to be the solvent.[a]

A.2 UTILITY REQUIREMENTS

A.2.1 Calculation of Natural Gas Requirements

The natural gas requirements depend on the exhaust flow rate, waste gas heat content and temperature, and the amount of heat recovery used. The heat content of the waste gas depends on the VOC concentration and species. As VOC concentration increases, the natural gas requirements generally decrease. Likewise, as the waste gas temperature and percent heat recovery increase, the natural gas requirements generally decrease. The results of these calculations are presented in Figure A-4. The VOC destruction efficiency is assumed to be 95 percent.

For example, suppose for a system without heat recovery, the waste gas flow is 5000 SCFM, the waste gas temperature is 300°F, and the VOC concentration (assuming toluene) is 10 percent LEL. Then from Figure A-4, the natural gas required will be 73 SCFM natural gas/10000 SCFM exhaust gas. From Section A.1.4 the exhaust flow rate is 5024 SCFM. Therefore the natural gas requirement will be 37 SCFM.

A.2.2 Calculation of Electricity Requirements

Electricity requirements are also based on the exhaust gas flow rate determined from the heat and material balance calculations. The pressure

TABLE A-2. INCINERATOR TEMPERATURE CONSTRAINTS

Symbol	Temperature Location	Constraint	Notes
T_{wg}	waste gas	Depends on application	
T_a	combustion air	70°F	Assumed
T_c	catalyst inlet	600°F minimum	Temperature required to insure initial VOC oxidation.
T_o	catalyst outlet	900°F minimum 1200°F maximum	Temperatures required to insure reasonable VOC oxidation without catalyst damage.
T_s	stack	500°F minimum	Temperature required to preclude acid condensation in stack.

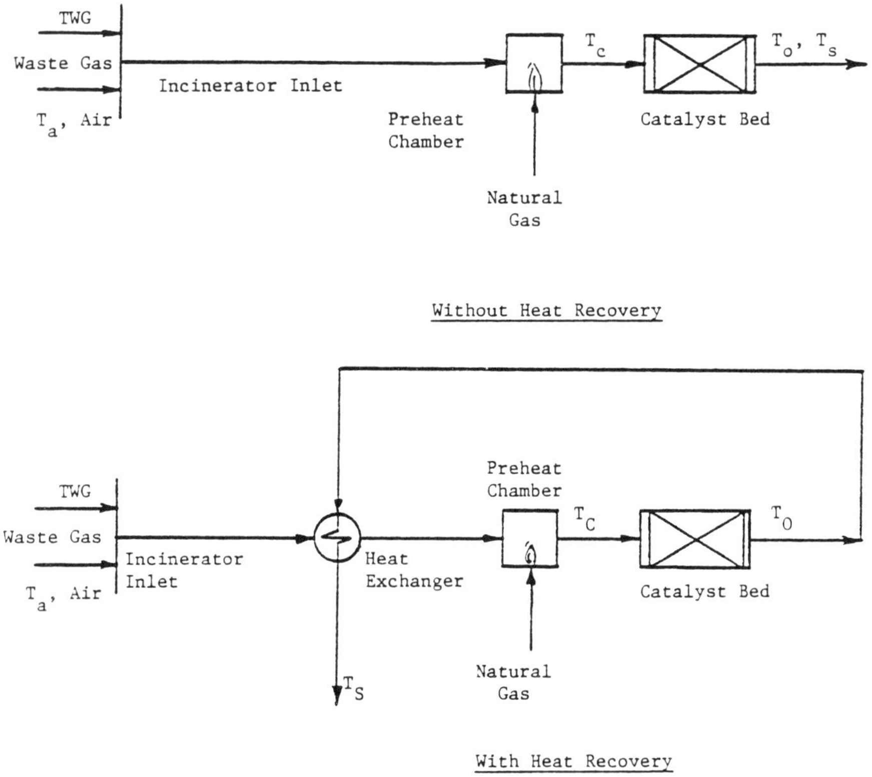

Figure A-1. Flow schemes for catalytic incinerators.

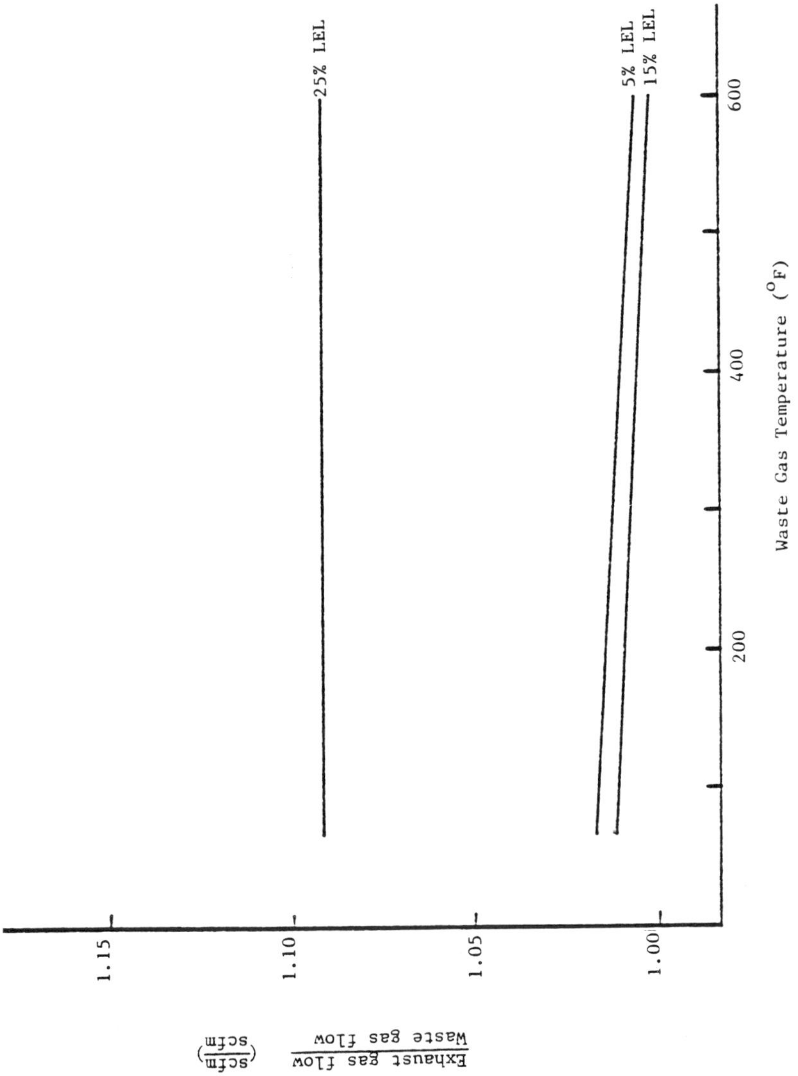

Figure A-2. Temperature and concentration effects upon $\frac{\text{exhaust gas flow}}{\text{waste gas flow}}$ ratio assuming no heat recovery.

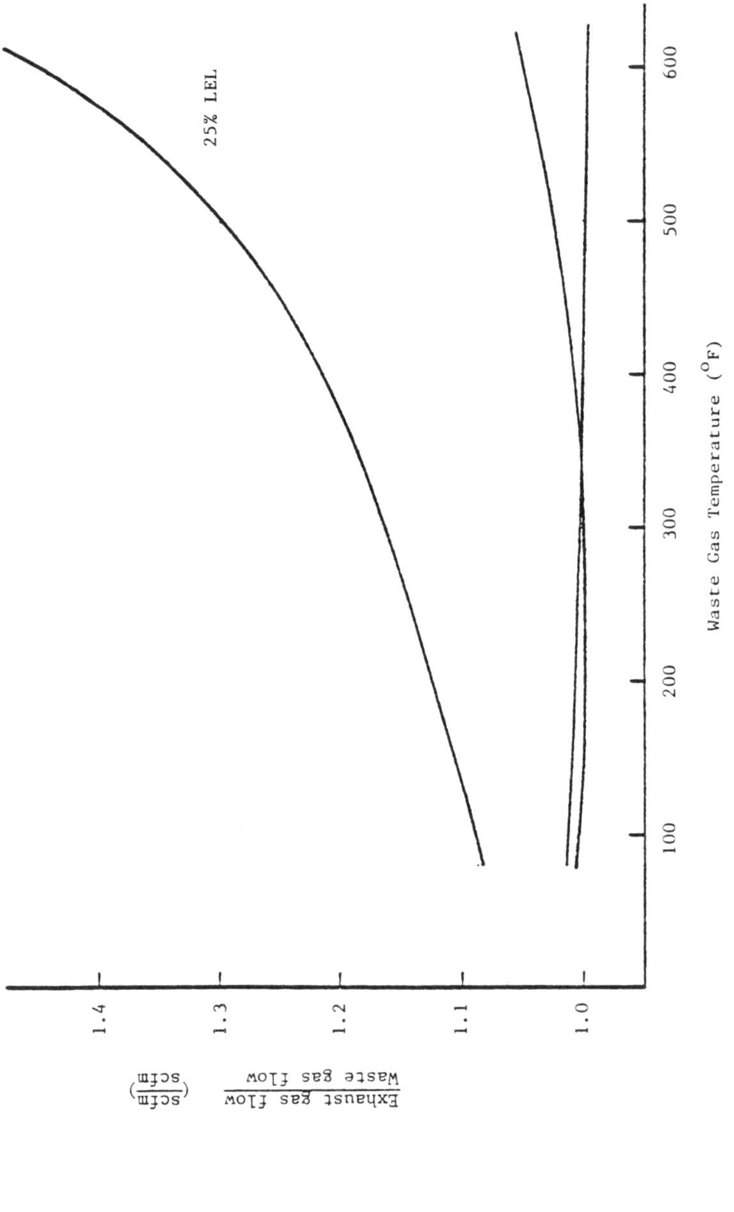

Figure A-3. Temperature and concentration effects upon $\frac{\text{exhaust gas flow}}{\text{waste gas flow}}$ ratio assuming 40% heat recovery.

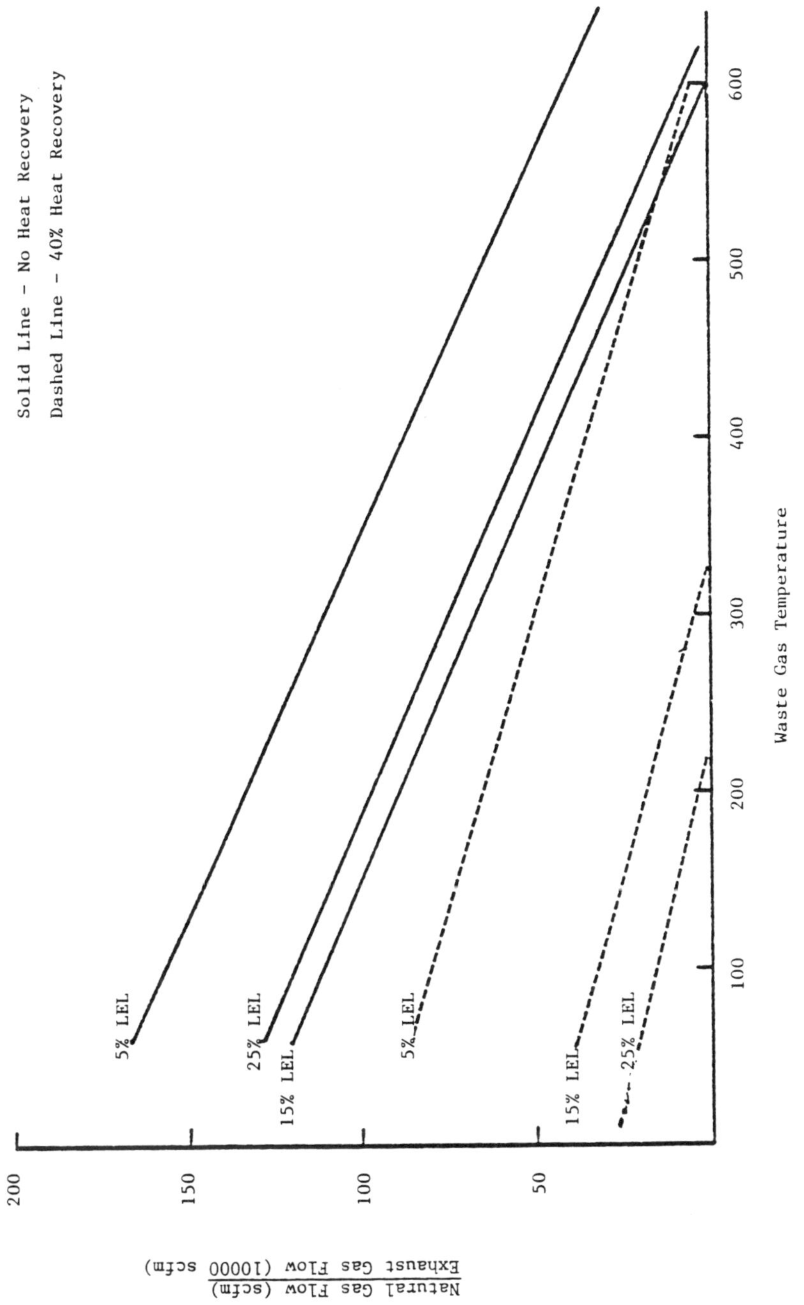

Figure A-4. Natural gas requirements.

drop across the incinerator was assumed to be 6" H_2O without heat exchange and 15" H_2O with heat exchange. A mechanical efficiency of 0.6 was assumed for the fan. Using these assumptions, the following factors were derived to provide the annual electricity requirements in kilowatt-hours given the exhaust flow rate:

$$\text{Exhaust flow rate (SCFM)} \times \frac{7.042\ \text{kWh}}{\text{yr SCFM}} = \frac{\text{kWh}}{\text{yr}} \quad \text{with no heat recovery.}$$

$$\text{Exhaust flow rate (SCFM)} \times \frac{17.605\ \text{kWh}}{\text{yr SCFM}} = \frac{\text{kWh}}{\text{yr}} \quad \text{with heat recovery.}$$

Appendix B

Identification of Plants Using Catalytic Incinerators

This Appendix summarizes the plants identified by this study that are using catalytic incinerators. According to one source, there are between 500 and 2,000 catalytic incinerators in use today. This study has identified at least 178. For most of the 149 plants identified as using catalytic incinerators, there is no information concerning the number of incinerators on site. In these cases, the number being used was assumed to be one. Therefore, for the plants identified, there are probably more than 178 catalytic incinerators. In fact, the average number of catalytic incinerators per site for those that specify that number is 3.1.

The plants are organized according to industry, and these industries are arranged according to the number of plants identified within each. Metal can coating tops the list with 25 plants. Magnet wire enameling is second with 23 plants, and so forth. The following information was provided for each plant when the information was available:

- company
- location
- type of process
- waste gas parameters
  - flow rate
  - temperature into incinerator
  - temperature out of incinerator
- VOC Concentration
- annual operating hours
- tons of VOC emitted per year
- performance data and method used to determine them
- other notes

The methods for determining VOC destruction efficiencies are assigned the acronyms STEM, MBE,, EF, and G. These have the following meanings:

- STEM (source test emission measurement) means that actual testing has been done.

- MBEK (material balance and engineering knowledge) means that efficiency was determined by material balances.

- EF (emission factor) means that a state or local agency has assigned a value for incinerators for which there is no data.

- G (guess) the value for efficiency was assigned a value by NEDS on the basis of a guess.

In the notes section, information about the number and types of incinerators on site and information about availability of heat recovery are included.

Plant identification and information were derived from the following sources:

- national emission data system (NEDS),

- state and local air quality control agencies,

- vendors, and

- literature.

Other data were obtained in a telephone survey of plants known to have emission test data.

The states that provided plant information included Georgia, Illinois, Indiana, Massachusetts, New Jeresy, Ohio, and Texas. The South Coast Air Quality Management District also supplied plant information.

Some catalytic incinerator vendors also provided information. They include DuPont, Met-Pro (formerly Oxycatalyst), Air Resources, and Englehard.

Literature sources include reports by Hydroscience and Kirk Otthmer's Encyclopedia of Chemical Technology.

Industry	Company	Location	Process	Waste Gas Parameters				Annual Operating Hours	Tons VOC emitted per year	Performance Data (%) and efficiency method	Other Notes
				Flow (scfm)	Temp(°F) inlet	Temp(°F) outlet	VOC concentration				
Metal Can Coating (Cont.)	Crown, Cork & Seal Co.	Faribault, MN	--	10100	-	300	--	--	13	90.0 EF	
				17900	-	130	--	--	13		
				42400	-	100	--	5200	13		
				44000	-	80	--	2100	40		
		Hudson County, NJ	--	7900	-	750	--	6500	11	90.0 MBEK	
		Philadelphia, PA	Coater and ovens	5900	-	350	--	--	--	95.0 EF	
				3500	-	300	--	6000	--	95.0 EF	
				2900	-	77	--	--	--	95.0 EF	
		Houston, TX	--	--	-	-	--	--	--	--	
	Kaiser Aluminum	Edison, NJ	--	--	-	-	--	--	--	--	DuPont Torvex
	National Can Corporation	La Puente, CA	--	--	-	-	--	4200	99	99.0 MBEK	
		Etiwanda, CA	Ovens	2000-19000	-	-	2100 ppm expressed as C	7500	11	90-94 STEM	8 incinerators natural gas fired, have heat exchangers
		Danbury, CT	--	20500	-	200	--	6200	--	90.0 MBEK	
	Owens Illinois	Glasboro, NJ	Ovens	--	-	-	--	--	--	--	1 DuPont Torvex, partial recycle to ovens
	Reynolds Aluminum	Torrance, CA	--	--	-	-	Solvent is toluol butyl acetate	8740	27	90.0 MBEK	
	Western Metal	El Segundo, CA	Ovens	1500	330	960-1060	9000-14000 ppm as CH_4	5600	6	96.0 STEM	
		San Francisco, CA	Ovens	3600	-	850	--	8740	-	45-85 EF	

Industry	Company	Location	Process	Waste Gas Parameters Flow (scfm)	Temp(°F) inlet	Temp(°F) outlet	VOC concentration	Annual Operating Hours	Tons VOC emitted per year	Performance Data (%) and efficiency method	Other Notes
Magnet Wire Enameling	Anaconda Industries	Green County, KY	Ovens	300	-	1050	--	8600	27	89.8 EF	
		Oldham County, KY	Ovens	300-2000	-	520-840	--	8740	--	85.0 EF	
		Sycamore, IL	Ovens	500	-	600	--	4200	2	90.0 MBEK	
	Essex International	Fort Wayne, IN	Ovens	-	-	-	--	--	--	--	
		Rockford, IL	Ovens	600-700	-	-	--	--	--	--	
		Hillsdale, MI	Ovens	-	-	-	--	8740	641	5.0 EF	
		Vincennes, IN	Ovens	1600	-	600	--	8400	18	95.0 EF	
	General Electric	Shreveport, LA	Ovens	800	-	690	--	2000	--	87.0 STEM	4 catalyst beds
		Hickory, NC	Ovens	-	-	-	--	--	--	--	
		Schenectady, NY	Ovens	40-500	-	500-900	25% LEL cresols	7500	--	90.0 STEM	UOP supplies catalyst beds, some are internal and some are external
		Rome, GA	Ovens	600-700	-	500	--	7200	--	85-95 STEM	9 catalytic incinerators, all external to ovens, electrically heated
		Pittsfield, MA	Ovens	-	-	-	--	--	--	80.0 MBEK	
		Shelbyville, IN	Ovens	100-1100	-	1250	--	8100	131	85.0 MBEK	
		Fort Wayne, IN	Ovens	50-100	-	1200-1250	--	8100	31	85.0 EF	

Industry	Company	Location	Process	Waste Gas Parameters: Flow (scfm)	Temp(°F) inlet	Temp(°F) outlet	VOC concentration	Annual Operating Hours	Tons VOC emitted per year	Performance Data (%) and efficiency method	Other Notes
Magnet Wire Enameling (Cont.)	New Haven Wire & Cable	New Haven, CT	Ovens	400	-	920	--	8400	37	90.0 EF	
	Packard Electric Division of General Motors	Youngstown, OH	Ovens	9000	-	650	--	6000	89	80.0 STEM	8 incinerators for 8 ovens. 4 are internal to ovens.
	Phelps Dodge Magnet Wire	Hopkinsville, KY	Ovens	150-2600	-	300-900	--	8400	--	85 STEM & EF	
		Fort Wayne, IN	Ovens	300	-	900	--	8400	284	90.0 EF	
	REA Magnet Wire	Laurinburg, NC	Ovens	-	-	350	--	8200	136	85.0 MBEK	
		Lafayette, IN	Ovens	-	-	77	--	8100	187	85.0 EF	Has heat exchanger
		Fort Wayne, IN	Ovens	-	-	320	--	8100	442	85.0 MBEK	
	Viking Wire	Danbury, CT	Ovens	-	-	-	--	6200	--	--	
	Westinghouse Electric	Abingdon, VA	Ovens	600-1000	-	-	--	--	--	--	
		Hampton, SC	Ovens	11000	-	740	--	8740	45	80.0 MBEK	
Organic Chemical Manufacturing	Air Products & Chemicals	Pasadena, TX	--	--	-	-	--	--	--	--	
	American Cynamid		--	--	-	-	--	--	--	--	ARI Econo-abator
	Armak Company	Lyons, IL	--	--	-	-	--	--	--	86.5-95.0 MBEK	
	Chemetron Corp.	La Porte, TX	Chlorinated hydrocarbon destruction	--	-	-	--	--	--	100 STEM	ARI Econo-abator
	Chevron Chemical		--	--	-	-	--	--	--	--	ARI Econo-abator
	Diamond Shamrock	Deer Park, TX	Ethylene dichloride process vent	--	-	-	--	--	--	--	

Industry	Company	Location	Process	Waste Gas Parameters: Flow (scfm)	Temp(°F) inlet	Temp(°F) outlet	VOC concentration	Annual Operating Hours	Tons VOC emitted per year	Performance Data (%) and efficiency method	Other Notes
Organic Chemical Manufacturing (Cont.)	Dow Chemical	Freeport, TX	--	--	-	-	--	--	--	--	
	DuPont	Beaumont, TX	Acryloni-absorber vent gas	80000	100	750	1.2 wt %	--	--	42.5 STEM	Has heat exchange
	Fritzsche Dodge & Olcott	Clifton, NJ	Spray dryer miscellaneous storage	2500 -	- -	600 70	-- --	6000 6000	2 0	99.9 G 99.9 G	Has heat exchanger, Met-Pro system
	Goodyear Tire	Houston, TX	Butadiene fugitive losses	-	-	300	--	8740	4200	99.9 EF	
	Hilton-Davis Chemical Co.	Hamilton County, OH	Printing ink manufacturing	450	-	700	--	4300	1	99.0 MBEK	
	Hooker Chemical		--	-	-	-	--	--	--	--	ARI Econo-abator
	Jack Daniel Distributor	Lynchburg, TN	--	5800	-	1600	--	830	0	99.0 EF	
	Jefferson Chemical Company	Port Neches, TX	Ethylene oxide	46700	-	450	--	8740	168	99.1 MBEK	
	SC Johnson & Son	Racine, WI	Polymer manufacturing	-	-	-	--	4200	2	99.0 EF	
	Silicone Products Division of GE	Saratoga, NY	Production of low molecular weight methyl siloxanes	-	-	77	--	3700	1	99.0 STEM	
	Syngas Company	Deer Park, TX	--	-	-	-	--	--	--	--	
	Union Carbide	Atlanta, GA	Polyvinyl manufacturing	4200	-	180	--	6200	0	99.9 EF	
		S. Charleston, WV	Ethylene oxide, ethylene glycol blow off from air oxidation plants	68000	-	-	--	--	--	--	

Industry	Company	Location	Process	Waste Gas Parameters: Flow (scfm)	Temp(°F) inlet	Temp(°F) outlet	VOC concentration	Annual Operating Hours	Tons VOC emitted per year	Performance Data (%) and efficiency method	Other Notes
Organic Chemical Manufacturing (Cont.)	Unknown	--	formaldehyde	-	70	950	1000 ppm formaldehyde 1000 ppm CH_3OH 4000 ppm dimethylether 16000 ppm CO	--	--	99.0 STEM	Engelhard incineration system
Printing Operations	Ambassador College	Los Angeles, CA	--	1400	-	900	4000 ppm as CH_4	4700	--	90.5 STEM	Oxycatalyst catalytic incinerator, no heat exchange
	Arcata Graphics	Los Angeles, CA	--	-	-	-	--	3600	3	90.0 MBEK	Has heat exchanger
	Arnold Print Works	Adams, MA	--	-	-	-	--	--	-	--	1 incinerator
	Dayton Press Inc.	Dayton, OH	--	7100	-	420	--	5700	1	90.3 MBEK	Has heat exchanger
	General Offset Printing Co.	Springfield, MA	--	1500	700	1000	664 ppm	--	-	95.0 STEM	2 incinerators
	Judd & Detweiler	Washington, DC	--	3500	-	540	--	3100	3	85.6 STEM	1 catalytic incinerator with heat recovery, drying system supplied by TEC (catalytic incineration integral part)
	Medallion Graphics	Los Angeles, CA	--	2300	170	850	700 ppm as CH_4	--	--	86.0 STEM	Matthey Bishop catalytic incinerator, no heat exchanger
	Pacific Press	Vernon, CA	Printing press	8000	280	810	700 ppm	--	--	94.5 STEM	
	Review & Herald	Washington, DC	--	3000	-	130	--	2100	--	--	Has heat exchanger

Industry	Company	Location	Process	Waste Gas Parameters: Flow (scfm)	Temp(°F) inlet	Temp(°F) outlet	VOC concentration	Annual Operating Hours	Tons VOC emitted per year	Performance Data (%) and efficiency method	Other Notes
Printing Operations (Cont.)	Rexham Corp.	Lancaster, SC	Roto-gravure printing press	--	-	-	--	6000	280	95.0 MBEK	Has heat exchanger
	RR Donnelley & Sons	Chicago, IL	--	2300-13900	-	700	--	3200	--	95.0 MBEK	
		Mattoon, IL	--	--	-	-	--	4300	12	80.0 MBEK	
		Middlesex County, CT	--	1800-2400	-	700	--	--	--	90.0 EF	
	Vi-Tex Packaging	Suffolk, VA	--	--	-	100	--	6000	145	65.0 MBEK	
	W.F. Hall Printing	Chicago, IL	--	--	-	-	--	--	--	--	Met-Pro catalytic incinerator
	World Color Press	Effingham, IL	--	--	-	-	--	--	--	--	
	XAS Color Inc.	Farmers Branch, TX	--	--	-	-	--	--	--	--	
Other Non-Metal Coating Operations	Champion Laboratories	Edward County, IL	Surface coating operation	-	-	-	--	4000	2	90.0 MBEK	Has heat exchanger
	Congoleum Corporation	Philadelphia, PA	Coating of linoleum	22700	-	275	--	5800	173	85.0 EF	
	Dura-Bernardin	Evansville, IN	Litho-graphic oven	-	-	800	--	6400	--	90.0 EF	
	Firestone Plastics	Pottstown, PA	Plastics coating	-	-	-	--	--	--	100 STEM	ARI Econo-Abator
	Kerr Glass Mfg.	Lancaster County, PA	Plastic coating	5000	-	150	--	2800	1	99.0 G	Met-Pro catalytic incinerator
	Pennwalt Corp.	Monmouth County, NJ	Dental Equip-ment & Supplies (Process oven)	37500	-	145	--	2000	9	95.0 MBEK	
	Richmond Tank Car Co.	Galveston, TX	Spray booth	23900	-	70	--	4200	--	28.0 EF	Under construction as of 1979
	Terbilt Motors Co.	Denton, TX	Heavy duty truck assembly	--	-	-	--	--		--	

Industry	Company	Location	Process	Waste Gas Parameters: Flow (scfm)	Temp(°F) inlet	Temp(°F) outlet	VOC concentration	Annual Operating Hours	Tons VOC emitted per year	Performance Data (%) and efficiency method	Other Notes
Other Metal Coating Operations	GAF Corporation	Johnson City, NY	Fabricated metal products	1300	-	98	--	5000	1	98.0 STEM	
	General Electric	Louisville, KY	Refrigeration and heating equipment manufacturing	35300	-	78	--	3700	--	90.0 EF	
	Mack Truck	Allentown, PA	Small parts coating oven	-	-	-	--	--	--	--	DuPont system
	Matthey Bishop	Wayne, PA	Coating of catalytic converters	6300	-	120	--	2200	13	96.5 STEM	Has heat exchanger
	Phoenix Closures	Chicago, IL	Metal stampings	2800	-	800	--	1600	--	95.0 MBEK	
	USDOE Oak Ridge Gaseous Diffusion	Oak Ridge, TN	Motor curing oven	-	-	-	--	--	31	67.0 EF	
Paper Coating (Including PST&L)	Avery Label Co.	Monrovia, CA	--	5300	-	400	--	6000	227	90.0 MBEK	
	BEE Chemical Co. Roll Leaf Div.	Fairfield, NJ	Drying ovens	5800	-	550	--	2100	8	95.0 G	With heat exchanger
	General Formulations	Sparta, MA	--	-	-	-	1000 ppm	5000	--	95.0 STEM	DuPont torvex
	International Label	Clarksville, TN	--	33700	-	400	--	8400	41	90.0 MBEK	
		New York City, NY	Lacquering of whiskey boxes	-	-	-	--	--	--	--	Englehard system
Asphalt Blowing	Chevron USA	Mobile, AL	--	5700	-	230	--	900	1	99.9 EF	
	Flintkote Co.	Atlanta, GA	--	8600	-	640	--	8740	2	99.0 G	
		St. Paul, MN	--	5800	-	600	--	6000	1	92.7 EF	
	GAP Corporation	Denver, CO	--	800-16000	-	200-350	-	8740	483	90.0 EF	

Industry	Company	Location	Process	Waste Gas Parameters: Flow (scfm)	Temp(°F) inlet	Temp(°F) outlet	VOC concentration	Annual Operating Hours	Tons VOC emitted per year	Performance Data (%) and efficiency method	Other Notes
Asphalt Blowing (Cont.)	Intec Inc.	Port Arthur, TN	--	-	-	-	--	--	--	--	
Alkyd Varnish and Resin Production	Bisonite Company	Tonawanda, NY	Kettles	50	-	900	--	800	1	95.0 STEM	Batch process, catalyst replaced every 14-16 months
	Cargill	Carpentersville, IL	Varnish cook	-	-	-	--	--	--	--	ARI-Econo-Abator
	Desoto, Inc.	Garland, TX	Storage of paints, varnishes, and lacquers	-	-	77	--	8740	5	99.0 EF	
	Perry & Derrick Paint Co.	Dayton, KY	Varnish cook	1600	-	650	--	500	0	98.0 EF	
Tire Manufacturing	B.F. Goodrich	Thomaston, GA	--	106000	-	290	--	7200	13	99.0 STEM	DuPont torvex
	Gates Rubber Co.	Denver, CO	Cord treating oven	2400-3800	500	-	--	6000	--	90.0 EF	DuPont torvex system, heat exchanger
	Goodyear Tire	Comanche County, OK	--	4700	-	70	--	8740	296	80.0 MBEK	Has heat exchanger
	Uniroyal Inc.	Mishawaka, IN	Photo-chemical printing polymer and ex-panded vinyl & rubber spreader	3600-5100	-	95	2800-13300 ppm as CH_4, 1/3 MEK, 2/3 toluene	7700	250	90.6 STEM	4 Met-Pro incinerators, single pass heat exchanger with 45% efficiency
Metal Production	Acco American	York County, PA	Ipsen tonpering furnace	340	-	850	--	6000	2	90.0 EF	
	Crucible Steel Casting	Cuyahoga County, OH	Core drying ovens	900	-	950	--	2400	2	90.0 G	
	Forest City Foundries	Cuyahoga County, OH	Core drying ovens	3500	-	300	--	2000	1	99.0 MBEK	
	Turner Seymour Manufacturing	Litchfield Co., CT	Cupola	800	-	2700	--	--	--	--	

Industry	Company	Location	Process	Waste Gas Parameters Flow (scfm)	Temp(°F) inlet	Temp(°F) outlet	VOC concentration	Annual Operating Hours	Tons VOC emitted per year	Performance Data (%) and efficiency method	Other Notes
Fabric Coating	Chemical Fabrics Corporation	Bennington, VT	Teflon sinter oven	2700	-	190	--	6500	--	85.0 EF	Has heat exchanger, DuPont torvex system
	DuPont	Fairfield, CT	--	-	-	-	Chemicals in waste gas are urethane, silicones, and neoprenes	--	--	--	1 incinerator, using urethane
	Wolverine Fabrics	Blacksburg, VA	--	8000	500-600	900	--	6000	67	93.0 STEM	Torvex catalytic incineration systems, no heat exchange
Metal Coil Coating	Alsco Anaconda	Zanesville, OH	--	5900	-	1400	--	6000	4	99.0 MBEK	
	Crown, Cork & Seal	San Francisco, CA	--	-	-	-	--	--	--	--	
	Kaiser Aluminum	Toledo, OH	--	20000	580	780	600-4500 ppm	--	--	99.5 STEM	1 catalytic incinerator, supplied by DuPont, no heat exchange
Miscellaneous Operations	Airco Spear Carbon	Youngstown, PA	Bake ovens for carbon and graphite products	1500	-	500	--	8740	26	90.0 G	Has heat exchanger
	Amoco	Fruitvale, TN	--	--	-	-	--	--	--	--	
	Ashland Oil Company	Beaver County, PA	--	900	-	400	--	8400	1	99.0 EF	Has heat exchanger
	Cabot Corp.		--	--	-	-	--	--	--	--	ARI Econo-abator
	Dresser Industries	Wharton, TX	--	--	-	-	--	--	--	--	
	E.I. DuPont Experimental Station	Wilmington, DE	Research and development lab	600-	-	72-	--	6000	--	65-90 MBEK	
	El Paso Natural Gas		--	--	-	-	--	--	--	--	ARI Econo-abator
	Exxon USA		--	--	-	-	--	--	--	--	ARI Econo-abator
	Fletcher Oil & Refining		--	--	-	-	--	--	--	--	ARI Econo-abator
	Gulf Oil	Crystal Beach, TX	--	--	-	-	--	--	--	--	

Part II

Case Studies

The information in Part II is from *Control of Industrial VOC Emissions by Catalytic Incineration: Volume 2. Final Report on Catalytic Incinerator Performance at Six Industrial Sites* by M.A. Palazzolo, R.M. Parks, and K.K. Fidler of Radian Corporation for the Office of Research and Development, U.S. Environmental Protection Agency, July 1984.

1. Introduction

Radian Corporation has conducted for the U.S. Environmental Protection Agency a testing program (Contract No. 68-02-3171, Work Assignment 50) to document the performance of catalytic incinerators applied to industrial processes for volatile organic compound (VOC) control. This final report summarizes the results of the testing program and presents a comparative analysis of results for all the incinerators tested.

VOC emissions are of concern to EPA because they may react photochemically in the atmosphere to form photochemical oxidants. National Ambient Air Quality Standards (NAAQS) for photochemical oxidants are routinely exceeded in many areas of the United States. The objectives of the EPA program under which this work was done are two-fold:

(1) to provide an overview of how catalytic incineration is applied to control industrial VOC emissions and to access the overall performance typically achieved in various applications, and

(2) to gather actual performance data through an EPA sponsored testing program on operating industrial catalytic incinerators.

The first objective was addressed in an earlier study done by Radian Corporation and the results are contained in Volume 1 of this report, entitled "Assessment of Catalytic Incineration and Competing Controls." The second objective was accomplished in the current testing program. The results of the testing program are contained in this Volume 2 final report and a series of six site-specific test reports (Volumes 3-8). The site-specific reports are hereafter referred to in this final report as "Site Test Reports." These test reports contain detailed information on the incinerator tests at each of the industrial sites.

The approach taken to gather test data on industrial catalytic incinerators consisted of (1) developing plant-specific test plans and test methods to characterize incinerator performance and (2) performing several days of comprehensive on-site testing on each incinerator. The testing focused on simultaneous monitoring of inlet and outlet VOC concentrations and measurement of incinerator operating conditions. This testing resulted in a data base on catalytic incineration that:

- is well documented in terms of incinerator and process parameters,
- uses the same testing methods on a number of catalytic incinerators applied in several industries, and
- provides continuous incinerator performance data over several days and numerous conditions.

The data contained in this report accurately represent the performance of the incinerators tested. It should be noted that other incinerators applied to similar processes may achieve better or poorer performance, depending on site-specific factors such as incinerator design and operation. The effect of major design and operating parameters on performance are discussed in this and other volumes of this report.

Section 2.0 of this report contains a summary of the findings of the test program while Section 3.0 discusses the criteria used to select the plant sites that were tested. Section 4.0 contains process descriptions of all the plants tested. Test procedures used during the course of the program are discussed in Section 5.0 and the test results are discussed in Section 6.0. Recommendations from the test effort are included in Section 7.0.

2. Summary

A total of eight catalytic incinerators were tested at six industrial sites between November 1982 and March 1983. Selection of the test sites was based on several criteria including process type and incinerator design characteristics.

All eight incinerators were used to control volatile organic compound (VOC) emissions from solvent evaporation processes. In these processes, pigments, inks or resins dissolved in organic solvents are applied to metal or paper surfaces. The solvents are then driven off the surfaces in drying ovens and the oven exhaust is ducted to a catalytic incinerator for solvent destruction. Incinerators at can coating, coil coating, magnet wire and graphic arts printing plants were tested.

Incinerator performance was characterized in terms of destruction efficiency, outlet solvent concentration and energy usage. Inlet and outlet solvent concentrations were monitored with hydrocarbon analyzers during a nominal 1-week test period at each site. Incinerator design and operating data, such as operating temperature, solvent type, catalyst volume and catalyst age, were collected on each incinerator to document the operating conditions during the test.

Four of the eight incinerators showed destruction efficiencies in excess of 90 percent under typical plant and incinerator operating conditions. These incinerators were applied in the can coating, coil coating, and graphic arts printing industries. Destruction efficiencies for each incinerator generally varied over a narrow range for different coatings or inks.

A fifth incinerator showed slightly lower efficiencies between 88 and 94 percent. This incinerator was also applied in the can coating industry.

The three remaining incinerators all showed comparatively low destruction efficiencies of approximately 80 percent. The lower efficiencies at two of these sites is attributed to the catalyst condition. One of the catalyst beds was fouled or deactivated by high temperature operation or material masking and the other catalyst bed appears to have been deactivated by normal catalyst aging. The lower efficiency at the third site is due to a lower solvent laden air (SLA) residence time. This incinerator, which was a much older design, had a SLA residence time approximately half that of the newer incinerators tested.

Two of the three incinerators that showed comparatively low destruction efficiencies were applied in the graphic arts printing industry and the third incinerator was applied to a magnet wire coating line.

Outlet total hydrocarbon (THC) concentrations based on EPA Method 25A ranged widely for the eight incinerators from 46 to 1600 parts per million by volume carbon (ppmv). However, measured outlet concentrations were below 350 ppmv for six of the eight incinerators.

Energy usage for most of the incinerators ranged from 78 to 270 kJ/Nm^3 (2.1 to 7.2 Btu/scf) of waste gas treated. One incinerator had an estimated energy usage of 780 kJ/Nm^3 (21 Btu/scf).

Comparison of destruction efficiencies measured according to Method 25A and Method 18 showed good agreement.

At several test sites, the incinerator operating temperature was varied to observe the effect of temperature on performance. As expected, destruction efficiency increased with increasing operating temperature.

However, the effect of temperature on destruction efficiency varied widely for the different incinerators. Of five incinerators for which the temperature was varied, two showed a fairly small effect of temperature and three showed a comparatively large effect. The largest increase in efficiency for a temperature rise of 14°C (25°F) was 2.5 percent, from 84.0 to 86.5.

Inlet concentration as measured by Method 25A was also observed to have an effect on destruction efficiency. For similar catalyst bed inlet temperatures, destruction efficiency increased with increasing inlet concentration. Variations in inlet concentration generally occurred for different coatings and coating rates. The effect of inlet concentration was found to be greatest for incinerators operating at efficiencies below 85 percent.

3. Test Site Selection Criteria

The criteria used to select host sites for the catalytic incineration test program are discussed in this section. These criteria were based on the data gaps and information needs identified in Volume I of this report, entitled "Control of Industrial VOC Emissions by Catalytic Incineration." All of the objectives of the EPA catalytic incineration program, which were discussed in Section 1.0, were considered and evaluated as the host sites were selected and tested.

The catalytic incinerator testing program was designed to examine catalytic incinerator performance from two perspectives: (1) how well the technology performed when applied to specific industries that typically used catalytic incineration and (2) how incinerator performance was related to a variety of process variables.

As reported in Volume 1, part of the catalytic incineration program involved identifying industrial sites that used catalytic incineration for VOC control. A summary of the sites identified is given in Table 3-1 by generic industry type. It is important to note that this list of plants is a random compilation of sites based on available data. As such, the relative number of plants identified in any industrial sector may be biased due to the availability of information in that particular sector.

TABLE 3-1. SUMMARY OF SITES IDENTIFIED

Industry	Number of Sites Identified
Metal can coating	25
Magnet wire enameling	24
Organic chemical manufacturing	20
Printing operations	17
Paper coating	5
Asphalt blowing	5
Alkyd varnish production	4
Tire manufacturing	4
Foundry core production	3
Fabric coating	3
Metal coil coating	3
Other metal coating operations	6
Other non-metal coating operations	8
Miscellaneous	21
TOTAL	149

The 149 sites listed in Table 3-1 formed the basis for the site selection task. In order to fairly access the performance of catalytic incinerators in representative industries, an attempt was made to contact sites that would provide an adequate cross-section of the major industries using catalytic incineration. Of the 149 sites identified in Table 3-1, 86 sites, or 58 percent are represented by four industries: metal can coating, magnetic wire production, organic chemical manufacturing and printing operations.

In an effort to test at least one incinerator in each of these four major industries, 8 to 10 plants were contacted in each category. Industry participation in the test program was strictly voluntary and several factors resulted in more plants being tested in some areas and none in others. Table 3-2 lists the final industry make-up of the plants included in the test program.

TABLE 3-2. SOURCE CATEGORIES TESTED

Industry	Number of Incinerators Tested	Number of Plants Represented
Metal Can Coating	3	2
Printing Operations	3	2
Magnetic Wire Enameling	1	1
Metal Coil Coating	1	1
Total	8	6

An organic chemical site was not selected and tested because the plants contacted were unwilling to have their units tested in the particular time frame of the testing program. This was due in part to the current depressed economic situation. Many plants were running at levels considerably below their standard operating ranges and some plants had actually ceased operations. In these cases, the plants made the decision not to test due to potential non-representative operation of the process and incinerators.

In some other industries the plants that were contacted indicated that production levels were so high that any disruption of plant operation could not be tolerated. In these cases, the plants were likewise not tested.

Another important consideration in selecting test sites was geographic density. An attempt was made to choose sites in areas where other potential sites existed so that travel costs could be reduced. This attempt was not as successful as desired and was not held as a constraint on the site selection process.

In addition to testing a variety of industries, the testing program also evaluated incinerator performance as it relates to process variables.

There are many process variables affecting performance, and they can be organized into two categories: waste gas characteristics and catalytic incinerator operating parameters. These are presented in Table 3-3.

TABLE 3-3. FACTORS AFFECTING CATALYTIC INCINERATION PERFORMANCE

Waste Gas Characteristics
VOC concentration
temperature
flowrate
VOC species
Catalytic Incinerator Parameters
catalyst volume, age, and type
heat recovery configuration
operating temperature
age of incinerator
annual operating hours and consistency of operation
incinerator design
catalyst bed residence time

In organizing the test program, as many of these parameters as possible were considered. The extent to which the data base represents the possible range of operating variables in each industry category was limited by the number of sites tested. Nevertheless, test sites were successfully selected to provide a data base spanning a wide range of waste gas characteristics and incinerator operating conditions. These characteristics and conditions are presented and discussed in Section 6.1. Additional discussion of incinerator operating conditions at each site is contained in the Site Test Reports.

4. Process Descriptions

Plants C-1 through C-6 represent a variety of industrial processes which utilize catalytic incinerators to destroy VOC emissions. Process descriptions for Plants C-1 through C-6 are presented in the following subsections. Included in each process description is a brief description of the incinerators tested at each plant and typical operating parameters for the tested incinerators. More complete discussions of the various processes and incinerator operating parameters are included in the Site Test Reports for Plants C-1 through C-6.

4.1 PLANT C-1

Plant C-1 produces decorated metal sheets and falls under SIC 3411, the metal can industry. In this process, pigments or varnishes dissolved in solvents are applied to metal sheets in thin coats. The solvents are then driven off the metal sheets in drying ovens and the oven exhaust is ducted to a catalytic incinerator for solvent destruction.

Two incinerators, incinerators 27 and 28, were tested at Plant C-1. Each incinerator is used to destroy solvent emissions from a single sheet metal coating line. Fugitive VOC emissions from a roll-type coating applicator are captured with a hood and ducted to the incinerator with the exhaust from three oven zones.

Coatings used at Plant C-1 are purchased pre-mixed from coating manufactures. Major solvents identified in the coatings include methylisobutyl ketone (MIBK), toluene, xylenes and ethyl benzenes.

A schematic diagram of the relationship between each oven and incinerator at Plant C-1 is shown in Figure 4-1. Both incinerators tested have identical configurations.

As shown in Figure 4-1, solvent laden air (SLA) drawn from the oven by a large blower passes into the burner zone of the incinerator. The SLA is heated by the combustion of natural gas and then passes through the catalyst bed where solvents in the SLA are oxidized. Approximately 40 percent of the incinerator outlet gas, typically at 400°C (750°F), is recirculated to the three oven zones. Dampers control the amount of air recirculated to maintain a preset oven zone temperature. This zone temperature varies with the coating being dried. The rest of the incinerator outlet gas exhausts to the atmosphere via a stack.

The actual gas flowrate through the incinerators at Plant C-1 is relatively constant for different products. However, the inlet VOC mass flowrate and the VOC type vary from one product to the next.

Incinerator 28 handles a SLA flowrate of approximately 3.2 Nm^3/s at 89°C (6780 scfm at 190°F) and incinerator 27 handles 3.6 Nm^3/s at 121°C (7,600 scfm at 250°F). Based on THC measurements, inlet solvent mass flowrates for the two incinerators ranged from 13 to 44 kg/hr (29 to 97 lb/hr), depending on the product being made. Actual solvent usage rates were not monitored during the testing.

The incinerators tested at Plant C-1 were designed and manufactured by DuPont and had a destruction efficiency guarantee of 90 percent. The catalyst in the incinerators is a catalyst coated ceramic honeycomb with the trade name Torvex^R. Each incinerator contains 0.33 m^3 (12 ft^3) of catalyst, giving a SLA residence time of approximately 0.03 seconds.

Incinerator 28 was started up in December 1980 and the original catalyst was replaced on November 13, 1982. Incinerator 27 began operation in March 1982 and still had the original catalyst when the plant was tested in November 1982.

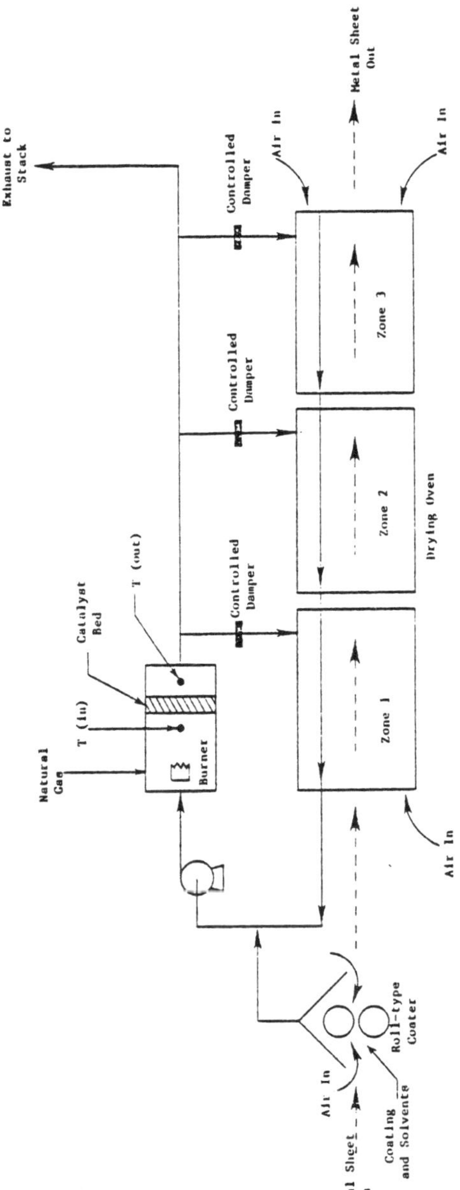

Figure 4-1. Process flow diagram for Plant C-1.

4.2 PLANT C-2

Plant C-2 produces graphic arts (magazines, brochures, etc.) by web offset printing. The inks used in the printing consist of pigments borne in a heavy organic solvent of C_{12} through C_{18} hydrocarbons. After the ink is applied to the paper web, the solvent is driven off in a gas-fired drying oven, leaving the pigments on the web. Oven exhaust gas is ducted to a catalytic incinerator for solvent destruction.

The incinerator tested at Plant C-2 is used to destroy solvent emissions from the drying oven on a single printing line. In addition to the drying oven exhaust gas, fugitive VOC emissions from the cooling zone immediately following the drying oven are captured with a hood and ducted to the incinerator.

A flow diagram illustrating the oven and incinerator configuration at Plant C-2 is given in Figure 4-2. As shown in Figure 4-2, solvent laden air (SLA) is drawn from the oven and cooling zone by a blower and ducted to the burner zone of the incinerator. The SLA is heated by the combustion of natural gas and then passes through the catalyst bed where solvents are oxidized.

The incinerator at Plant C-2 handles a SLA flowrate of approximately 0.9 Nm^3/s (2,000 scfm) at 101°C (215°F)and was designed by Matthey-Bishop with a design destruction efficiency of 90 percent. The ceramic catalyst in the incinerator is arranged in a honeycomb configuration. The total catalyst volume is 0.09 m^3 (3 ft^3), giving a SLA residence time of approximately 0.02 seconds.

The incinerator at Plant C-2 was started up in 1975. Initially the incinerator was believed to have logged only 2000 hours of operating time prior to testing in December 1982. This was later found to be considerably in error and the actual number of operating hours logged is not known.

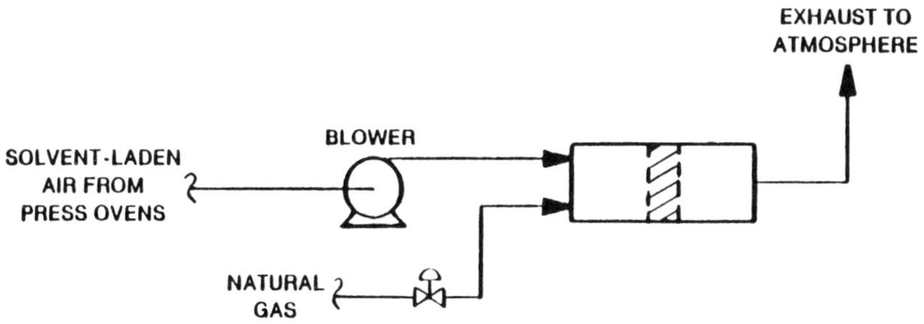

Figure 4-2. Plant C-2 catalytic incinerator flow diagram.

4.3 PLANT C-3

Plant C-3 produces graphic arts, primarily magazines and brochures, by web offset printing. Inks consisting of pigments borne in a heavy organic solvent of C_{12} through C_{18} hydrocarbons are applied to a paper web. The printed web passes through a drying oven where the solvent is evaporated. Oven exhaust gas containing vaporized solvent is ducted to a catalytic incinerator for VOC control.

Two incinerators, incinerators 20 and 21, were tested at Plant C-3. Incinerator 20 handles oven exhaust from a single printing line while incinerator 21 handles oven exhaust from two printing lines. The ink applied by the presses consists of colored pigments borne in a heavy organic solvent consisting of C_{12} through C_{18} organics with an average molecular weight of approximately 210 g/g-mole. Exact composition of the solvent is unknown.

Incinerators 20 and 21 are described separately below.

Incinerator 20--

A schematic diagram of the relationship between the oven and incinerator on printing line 20 is given in Figure 4-3. As shown in Figure 4-3, oven exhaust containing vaporized solvent is delivered by a single blower to the incinerator. The SLA is preheated by an integral heat exchanger which extracts heat from the incinerator exhaust gas, and then passes to the burner zone of the incinerator. The SLA is further heated by the combustion of natural gas and then passes through the catalyst bed where solvents in the SLA are oxidized. The combustion products are then cooled in the SLA preheater and exhausted to the atmosphere. Heat recovery is estimated to be 35 percent.

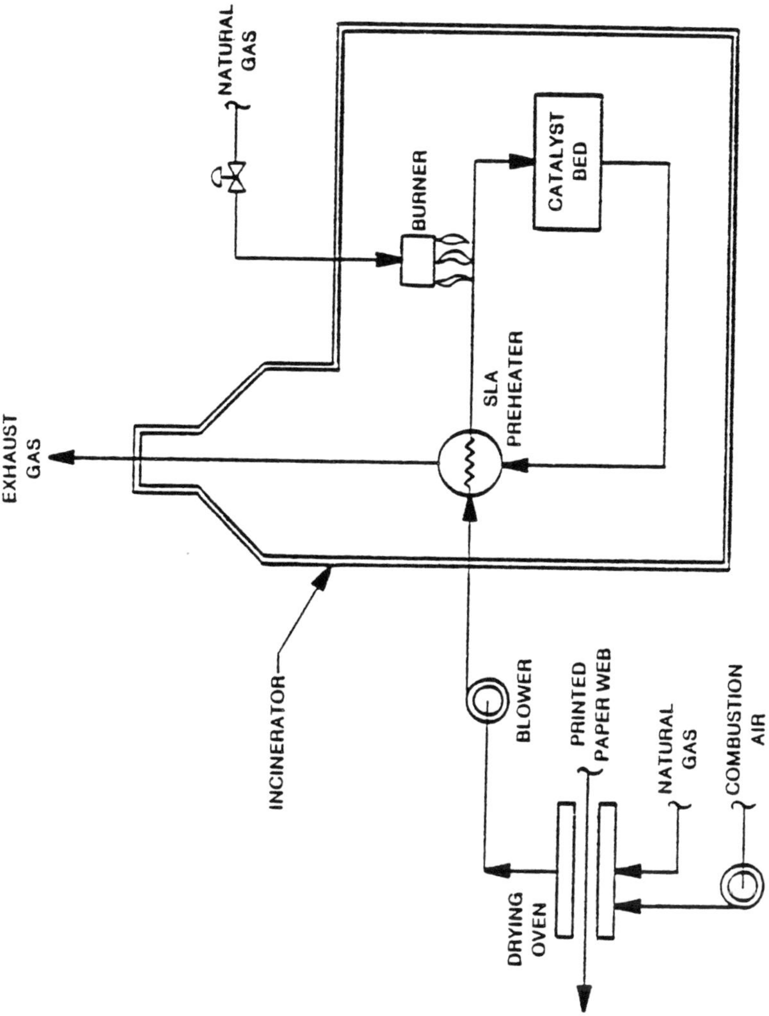

Figure 4-3. Process flow diagram for Incinerator 20 - Plant C-3.

Incinerator 20 handles a SLA flowrate of approximately 1.3 Nm^3/sec at 177°C (2,660 scfm at 350°F). The incinerator was designed and manufactured by TEC Systems. The catalyst consists of alumina-based pellets coated with oxides of copper, manganese, and trace amounts of palladium. The total catalyst volume is 0.302 m^3 (10.7 ft^3). Assuming a void fraction within the bed of 40 percent, the SLA residence time is 0.04 seconds.

Incinerator 20 was started up in February 1980 and the original catalyst was replaced in September 1981. The incinerator was tested in January 1983.

Incinerator 21--

A schematic diagram of the oven and incinerator on printing line 21 is shown in Figure 4-4. As shown in Figure 4-4, pressroom air is drawn into the system by a blower and is preheated in a heat exchanger which extracts heat from the incinerator exhaust gas. The heated air enters the two gas-fired drying ovens (press 21 has dual printing lines) and evaporates solvent on the printed web. The SLA is then further heated in a second heat exchanger and by the combustion of natural gas in the burner zone upstream of the catalyst chamber. The incinerator exhaust gas then exits the catalyst chamber, is cooled in the two heat exchangers mentioned above, and is discharged to the atmosphere. Heat recovery is estimated to be 60 percent.

Incinerator 21 handles a SLA flowrate of approximately 2.2 Nm^3/s at 178°C (4,670 scfm at 352°F) and was designed and manufactured by TEC Systems. The catalyst consists of.alumina-based pellets coated with oxides of copper, manganese, and trace amounts of palladium. The total catalyst volume is 0.491 m^3 (17.3 ft^3). Assuming a void fraction within the bed of 40 percent, the SLA residence time is approximately 0.04 seconds.

Incinerator 21 was started up in April 1982 and the original catalyst was in place during the testing in January 1983.

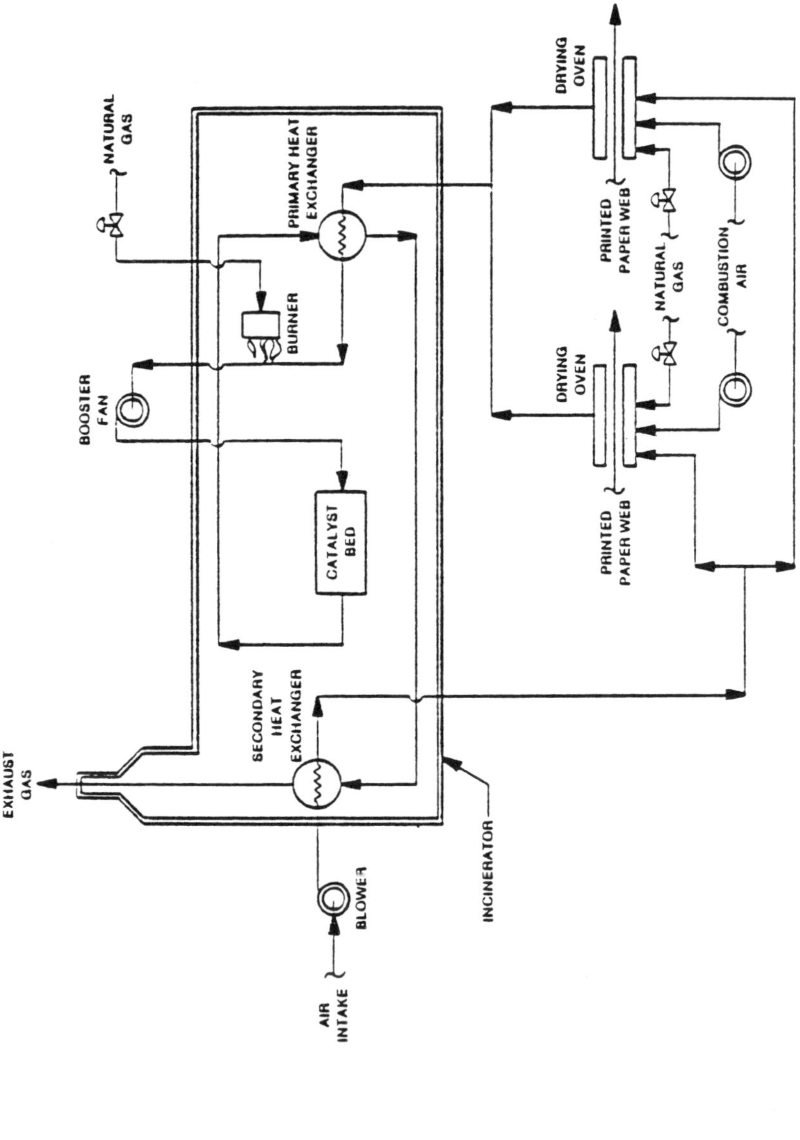

Figure 4-4. Process flow diagram for Incinerator 21 - Plant C-3.

4.4 PLANT C-4

Plant C-4 is a manufacturer of decorative metal cans and falls under SIC 3411, the metal can industry. In this process, coatings and pigments are applied to metal sheets in thin coats. The solvents are then driven off the metal sheets in drying ovens and the oven exhaust is ducted to a catalytic incinerator for solvent destruction.

The incinerator tested at Plant C-4 is used to destroy solvent emissions from a single sheet metal coating line. Coated metal sheets pass through five oven zones in addition to one preheat zone. Volatilized solvents are drawn from the various oven zones by a large fan through a common duct to the incinerator. In addition, fugitive VOC emissions resulting from the roll-type coating applicator operation are captured by a hood and ducted to the incinerator.

Coatings used at Plant C-4 are purchased pre-mixed from coating manufacturers. Solvents identified in the coatings include methyl isobutyl ketone (MIBK), cellosolve, xylenes and ethyl benzenes.

A schematic diagram illustrating the relationship between the drying oven and incinerator at Plant C-4 is given in Figure 4-5. As shown in Figure 4-5, solvent laden air (SLA) drawn from the oven by a large blower passes through the burner zone of the incinerator. The SLA is heated by the combustion of natural gas and then passes through the catalyst bed where solvents in the SLA are oxidized. The incinerator outlet gas exhausts to the atmosphere through a stack.

The incinerator handles a SLA flowrate of approximately 2.5 Nm^3/s at 164°C (5,330 scfm at 327°F). Based on measured THC concentrations, the inlet solvent mass flowrate ranged from 11 to 34 kg/hr (25 to 75 lb/hr), depending on the product type and size. Actual solvent usage rates were not monitored during the testing.

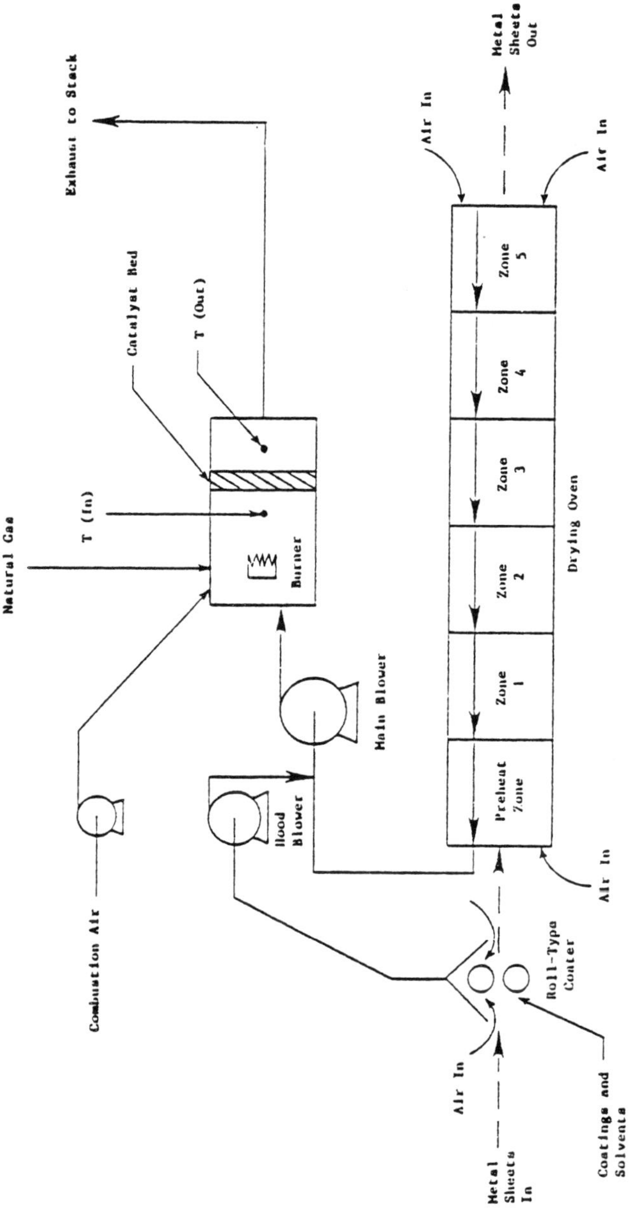

Figure 4-5. Process flow diagram for Plant C-4.

The incinerator at Plant C-4 is a Maxom fume burner retrofitted with a DuPont Torvex[R] honeycomb catalyst bed. The total catalyst volume is 0.28 m^3 (10 ft^3), giving a SLA residence time of approximately 0.03 second.

The incinerator at Plant C-4 was retrofitted with a catalyst bed in March 1982 and the original catalyst was in place during testing in February 1983. However, in September 1982, the catalyst was temporarily removed from the incinerator and cleaned with an alkaline solution. The catalyst had become masked from operation at a lower than optimum catalyst bed inlet temperature.

4.5 PLANT C-5

Plant C-5 produces magnet wire by coating bare copper wire with an electrically insulating enamel (SIC 33577). The wire is coated with a thin coat of resins dissolved in phenol and cresols. The coated wire passes through a drying oven where the solvents evaporate and the resins cure. The exhaust from the oven is ducted to a catalytic incinerator for solvent destruction.

A schematic diagram of the coating process and the catalytic incinerator at Plant C-5 is shown in Figure 4-6. As shown in Figure 4-6, solvent laden air (SLA) is drawn from the drying oven by a blower and passes into the burner zone of the incinerator. The gas is heated in the burner to approximately 399°C (750°F) by the combustion of a 9:1 air/fuel mixture. The hot gas then passes through the catalyst bed where the solvents are oxidized. All of the outlet gas from the incinerator is exhausted to the atmosphere via a stack.

The SLA gas flowrate at the incinerator inlet is approximately 0.28 Nm^3/s at 233°C (603 scfm at 450°F). The inlet gas flowrate remains constant during normal operation. However, the inlet solvent mass flowrate varies with the type of coating, the number of wires, the wire diameter and the line speed.

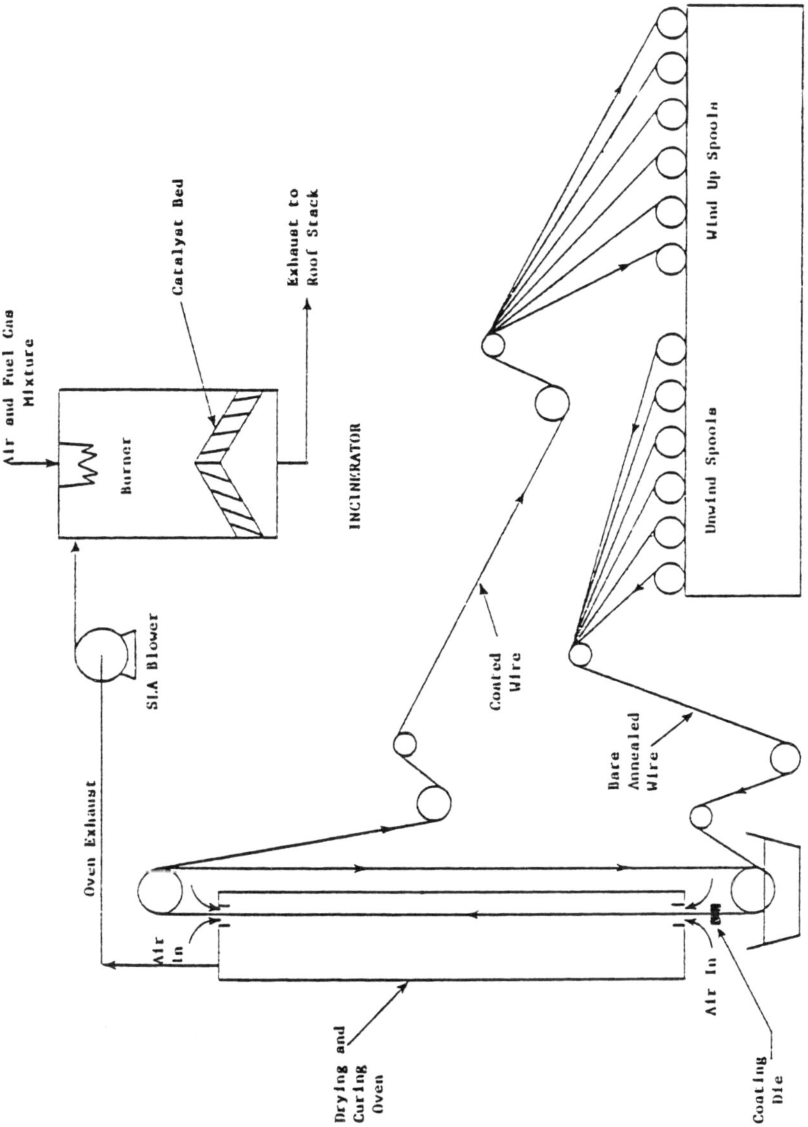

Figure 4-6. Process flow diagram for Plant C-5.

The incinerator tested at Plant C-5 was originally designed and manufactured by OxiCatalyst Incorporated (circa 1965) and was later modified by the plant. The modification involved switching from a ceramic-based catalyst to a metal-based catalyst. Further, the catalyst bed configuration was changed to increase the catalyst volume. Originally, the catalyst bed was perpendicular to the gas flow and not angled in the manner shown in Figure 4-6.

The catalyst tested consists of catalyst coated metal ribbon with a catalyst bed volume of 0.024 m^3 (0.83 ft^3). Assuming a void fraction of 40 percent, the SLA residence time is approximately 0.015 second. A recently rejuvinated catalyst bed was in place during the testing in February 1983.

4.6 PLANT C-6

Plant C-6 produces coated aluminum foil for use in the food packaging industry and falls under SIC 3497, the metal foil and leaf industry. In this coil coating process, aluminum foil is unwound from a coil and passed through a coating applicator where pigments dissolved in solvents are applied to the foil. The solvents are then driven off the foil in a drying oven and the coated foil is recoiled. The exhaust from the coating oven is ducted to a catalytic incinerator for solvent destruction.

The incinerator tested at Plant C-6 is used to destroy solvent emissions from a two-side coil coating line. The coating line is actually two coaters in series, with one coating station and drying oven for each side of the foil. For two-side coating, the foil must be passed through each coating station and its corresponding drying oven. Solvent emissions from only one of the coating stations and drying ovens are destroyed in the incinerator. Emissions from the other oven are currently not abated because this section of the oven is used infrequently.

In addition to the exhaust from the oven, fugitive VOC emissions from the coating area are also ducted to the incinerator. The fugitive emissions are captured with "floor sweeps" located on each side of the coating applicator.

The coatings used at Plant C-6 contain mostly toluene and methyl ethyl ketone (MEK). However, some of the coatings also contain xylenes, isopropyl alcohol, n-butyl alcohol and/or cellosolve acetate. Based on plant data, solvent usage rates averaged approximately 77 kg/hr (170 lb/hr) during the testing.

A process flow diagram showing the incinerator and coil coater is given in Figure 4-7. For simplicity, Figure 4-7 shows only one drying oven and coating applicator. Emissions from the other oven and coating applicator are not ducted to the incinerator tested.

As shown in Figure 4-7, a large blower draws solvent laden air (SLA) from the first two oven zones and from the floor sweeps. The SLA passes into the burner zone of the incinerator where it is heated to 288°C (550°F) by the combustion of natural gas. A portion of the floor sweep gas serves as the combustion air for the burner.

The hot gas from the burner zone passes first through a high temperature filter for particulate removal and then through the catalyst bed for solvent destruction. Approximately 50 percent of the incinerator outlet gas, typically at 399 to 427°C (750 to 800°F), and a portion of the floor sweep gas are recirculated to the first two zones of the oven. Dampers control the amount of air recirculated to maintain a preset oven zone temperature. This zone temperature varies with the coating being dried. The rest of the incinerator outlet gas exhausts to the atmosphere via a stack. Heat recovery is estimated to range from 25 to 36 percent, depending on the operating temperature.

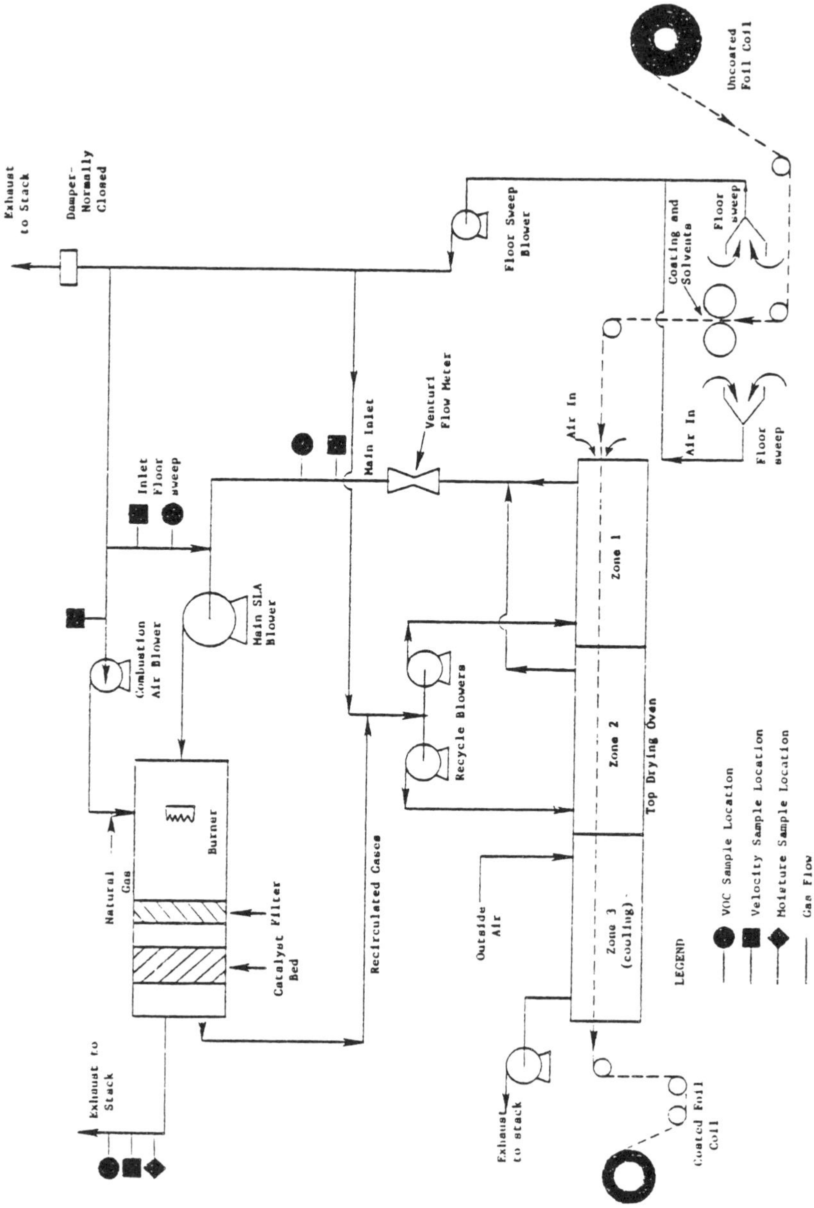

Figure 4-7. Process flow diagram for Plant C-6.

The cooling zone of the oven uses outside ambient air to cool the coated foil before it is recoiled. The exhaust from the cooling zone of the oven is vented back outside and does not pass through the incinerator.

The gas flowrate of SLA from the drying oven is constant for different coatings or products. The inlet solvent concentration and the solvent type vary from one product to the next.

The incinerator at Plant C-6 handles an inlet SLA flowrate of approximately 5.33 m^3/s at 149°C (11,300 scfm at 300°F). The incinerator was designed and manufactured by DuPont and had a destruction efficiency guarantee of 95 percent. The catalyst in the incinerator is a catalyst coated ceramic honeycomb with the trade name Torvex^R. The total catalyst volume is 0.50 m^3 (17.5 ft^3), giving a SLA residence time of approximately 0.03 seconds.

The incinerator began operation on January 24, 1983 and had the original catalyst at the time of testing in March 1983.

5. Testing Procedures

Testing procedures followed during the performance evaluations of catalytic incinerators at Plants C-1 through C-6 are presented in this section. In most cases, sampling and analytical procedures were based on EPA reference methods. However, in some instances, non-reference methods or modifications of reference methods were needed. Section 5.1 describes the various sampling and analytical methodologies used to evaluate the performance of catalytic incinerators at Plants C-1 through C-6. In Section 5.2, sampling and analytical procedures matrices are presented for Plants C-1 through C-6.

5.1 SAMPLING AND ANALYTICAL METHODS

In order to characterize operation and volatile organic compound (VOC) destruction efficiency of the incinerators at Plants C-1 through C-6, the following analyses were performed:

- VOC analyses
- Carbon monoxide (CO) analyses
- Fixed gas analysis (for molecular weight determination)
- Moisture content determination
- Velocity/volumetric flowrate determination

In addition, other process data including operating temperatures, pressures, fuel usage, etc., were measured or monitored during testing to further characterize incinerator operation. The various sampling and analytical methodologies used for each analysis are briefly described in the following subsections. A more detailed discussion of sampling and analytical procedures are found in the Interim Test Reports for Plants C-1 through C-6.

5.1.1 VOC Analyses

Three analytical procedures were used at Plants C-1 through C-6 to measure VOC concentration and to identify solvent components in the incinerator inlet and outlet gas streams. Total hydrocarbon (THC) analyses were performed on each incinerator based on EPA Proposed Reference Method 25A[1]. This was the primary method used to determine incinerator destruction efficiency. In addition, speciation of the inlet and outlet gas streams by gas chromatography (GC) was performed according to EPA Draft Reference Method 18[2]. Finally, samples for gas chromatography/mass spectrometry (GC/MS) analyses were taken at each incinerator tested to identify solvent components in the sampled gas streams.

Sampling and analytical procedures used in each analysis are described below.

Method 25A Analyses--

Both the incinerator inlet and outlet gas streams were continuously sampled. A three-hole sampling rake was used to extract the gas sample from the main gas stream. As required by Method 25A specifications, the sampling rake had holes 3.2 mm (1.8 in) in diameter drilled in the tubing at a relative location of 16.7, 50 and 83.3 percent of the equivalent duct diameter. Sample gas passed through 6 feet of heat-traced stainless steel line and a heated (205°C, 400°F) sampling pump before entering a heat-traced Teflon[R] umbilical that carried it to the mobile laboratory. During transport through the umbilical from the duct to the laboratory, the sample gas was maintained at a sufficiently high temperature to prevent condensation of organic components. Required umbilical temperatures varied for each incinerator depending on the composition of the sampled gas streams.

A manifold system inside the mobile laboratory was used to split the sample streams for analysis in two types of instruments, a THC analyzer and an organic species analyzer. A schematic of the typical sampling arrangement for measurement of VOC concentration is shown in Figure 5-1.

Method 25A specifies that THC determinations be performed using a continuous monitor equipped with a flame ionization detector (FID). Although an FID was used for the testing, monitoring was performed on a semi-continuous basis. The FID response was calibrated against propane standards. Three different instruments were used to measure THC concentrations in the incinerator inlet and outlet gas streams.

In most cases, inlet gas stream samples were analyzed by a Byron 301, THC analyzer. The Byron Model 301 Hydrocarbon Analyzer is a non-specific process gas chromatograph (GC) utilizing an FID to quantitatively determine methane (CH_4) and non-methane hydrocarbons in the vapor phase. In cases where the incinerator inlet stream contained high concentrations of high molecular weight organic compounds, a modified Tracor 560 gas chromatograph was used. In order to convert the Tracor 560 GC to an automated THC analyzer, the GC was modified to introduce gas samples directly to the FID and to inject samples automatically at 10 minute intervals.

Incinerator outlet gas samples were analyzed by a Byron 401 THC analyzer. The Byron Model 401 Total Emissions Analyzer is similar to the model 301. The 401 has the additional capability of analyzing a sample stream for total hydrocarbons (THC), nonmethane hydrocarbons, carbon dioxide (CO_2), methane, and carbon monoxide (CO). During the testing, the instrument was operated mostly in the THC only mode because the one minute cycle time allowed measurement of relatively rapid concentration changes. The full scan mode was operated for brief periods to obtain measurements of CO and CH_4 in the outlet gas.

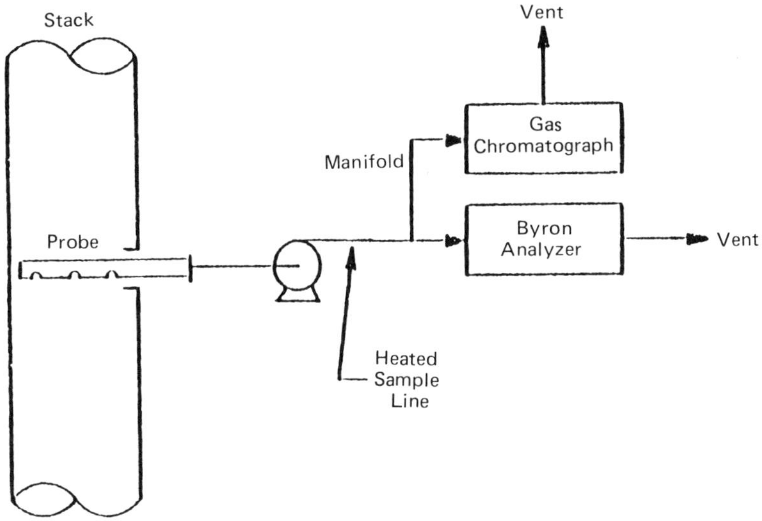

Figure 5-1. Sample acquisition system for VOC analyses.

Method 18 Analysis--

Sample acquisition for GC analysis was performed by the same sampling system used for THC analysis. As discussed, a manifold system inside the mobile laboratory was used to split the inlet and outlet sample streams for THC and GC analysis.

Draft Method 18 provides for the individual quantitation of organic species by gas chromatography using an FID. Two GC instruments were used during testing in a Hewlett Packard Model 5730A and an Antek Model 460. GC analyses were performed simultaneously with THC analyses wherever possible.

GC/MS Analysis--

Samples for GC/MS analysis were acquired from two sources. For some incinerators, samples for GC/MS analysis were acquired at the outlet of the sample acquisition system described above using evacuated canisters. The canisters were cleaned, evacuated, and the absolute pressure recorded before sampling. Immediately prior to sampling, canister pressure was remeasured to ensure that no leakage had occurred. Flow regulators were used to provide a constant flow over the 20-minute sampling period.

A second source of GC/MS samples at some incinerators was the impinger water from the moisture train. Hot outlet gas dissolves in the chilled impinger water, as evidenced by a color change. Samples of water from the first impinger were collected in clean sample bottles for GC/MS analysis.

All GC/MS analyses were performed by Radian's Austin laboratories. These analyses provided identification of various organic species in the sampled gas streams.

5.1.2 CO Analysis

CO concentrations of the incinerator exhaust gas were periodically measured by the Byron 401 THC Analyzer operating in the "Full Scan" (FS) mode. Sample acquisition occurred via the sampling system described in

Section 5.1.1. In the FS mode, a portion of the gas sample is introduced directly to the FID, yielding a THC measurement. Subsequently, nonmethane hydrocarbons (NMHC), CH_4, CO_2, and CO are chromatographically separated from the balance of the gas sample. One at a time, each of these groups is passed through an oxidizer/reducer to convert all carbon to CH_4. These reduced streams are then individually introduced to the FID for measurement, with a complete cycle requiring 12 minutes.

5.1.3 Fixed Gas Analysis

Two EPA Reference Method 3 equivalents were used to determine the dry (M_d) and wet (M_w) molecular weight of the incinerator inlet and outlet gas streams.[3]

In the first method, rather than collecting integrated or grab samples of the gas stream, the umbilical and manifold system described in Section 5.1.1 was used to acquire samples for fixed gas (CO_2, O_2, and N_2). After the sample stream passed through the organic species analyzer, it entered a set of impingers to knock out moisture and heavy organics. The cool, dry gas then proceeded through a Fisher Model 1200 Gas Partitioner for fixed gas analysis. The Fisher Gas Analyzer employs dual thermal conductivity detectors to determine the concentrations of fixed gases in a sample stream.

The second method of fixed gas analysis required two instruments. An oxygen Fyrite apparatus was used to determine oxygen content of the incinerator outlet stream. Gas sample for oxygen analysis was drawn directly from the outlet sampling port. The carbon dioxide content of the outlet stream was measured using the Byron 401 analyzer in the full scan mode. Gas sample for CO_2 analysis was drawn from the Method 25A sampling manifold. The nitrogen content of the outlet stream was determined by difference. For both methods, the dry molecular weight of the inlet and outlet gas was calculated based on the concentrations of the fixed gases and converted to a wet basis using Method 4 moisture data.

5.1.4 Moisture Content Determination

The moisture content of the incinerator inlet and outlet gas streams was determined using the methodology prescribed in EPA Reference Method 4. Based on this method, a measured volume of particulate-free gas is pulled through an impinger train.[4] Figure 5-2 graphically illustrates the sampling train which was used for moisture determination. The quantity of water condensed in the impingers is determined gravimetrically and then related to the volume of gas sampled to determine the moisture content.

5.1.5 Velocity/Volumetric Flowrate Determination

The volumetric gas flowrate of the incinerator inlet and outlet streams was measured using procedures described in EPA Methods 1 and 2 velocity of the flue gas.[5,6] The average gas velocity was calculated from the average gas velocity pressure (ΔP), the average flue gas temperature, wet molecular weight, and the absolute pressure in the stack. Temperature and ΔP, profiles were obtained by traversing the stack.

Temperature profile and ΔP data were collected at each of the sampling points using a K-type thermocouple and an S-type pitot tube. A gauge oil manometer was used to measure the pressure drop (ΔP) across the S-type pitot. The daily barometric pressure was obtained from a local weather station and corrected for evaluation.

For some incinerators, the volumetric gas flowrate of the incinerator inlet could not be measured according to EPA Method 2 because no acceptable sampling location was available. The inlet gas flowrate was calculated using the measured outlet gas flowrate and estimated combustion air and fuel gas flowrates.

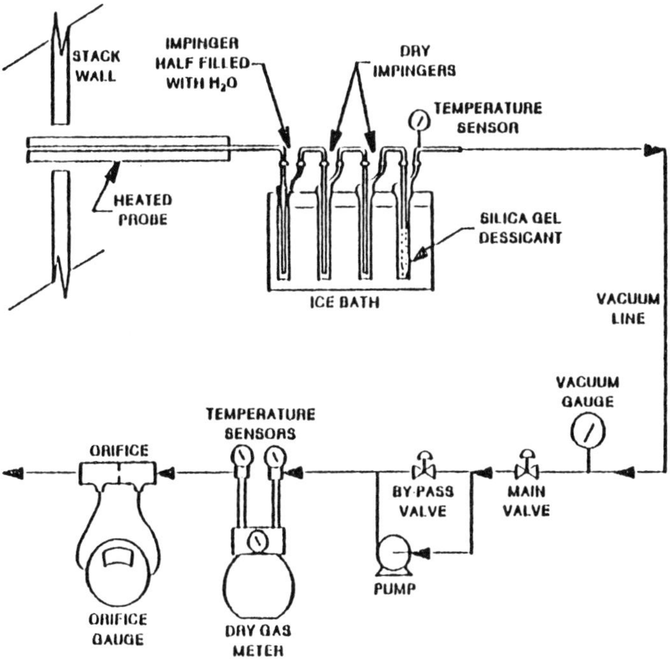

Figure 5-2. EPA Method 4 moisture train.

5.1.6 Other Process Data

In addition to VOC concentrations, CO concentrations, stream molecular weights and volumetric flowrates, several process parameters were measured or monitored at various points in the incinerators. These parameters included pressure drops across each stage of the system, operating temperatures, and combustion air and fuel parameters. In some instances, these data were obtained from plant instrumentation.

5.2 SAMPLING AND ANALYTICAL PROCEDURES MATRICES

A summary of the specific sampling and analytical procedures used during performance testing at Plants C-1 through C-6 are presented in Tables 5-1 to 5-6.

TABLE 5-1. SAMPLING AND ANALYTICAL PROCEDURES MATRIX - PLANT C-1

Measurement Parameter	Sampling Procedure	Analytical Procedure	Reference
VOC Concentration (THC)	Rake-type probe, heated sampling line (100°C (212°F))	FID-based THC analyzers	EPA Proposed Reference Method 25A
VOC Concentration (Speciated)	Rake-type probe, heated sampling line (100°C (212°F))	FID-based gas chromato-graph	EPA Draft Reference Method 18
VOC Species Identification	Rake-type probe, heated sampling line (100°C (212°F)), evacuated canisters	GC/MS	--
CO Concentration	Rake-type probe, heated sampling line (100°C (212°F))	FID-based analyzer with with species separation and oxidation/reduction to methane.	
Molecular Weight (Fixed Gases)	Rake-type probe, heated sampling line (100°C (212°F)	TCD-based analyzer	EPA Reference Method 3 (Equivalent)
Moisture Content	Goose-neck probe, chilled impinger train	Gravimetric determination of condensate, measurement of sampled gas volume	EPA Reference Method 4
Velocity/Volumetric Flow Rate	S-type pitot tube	Calculations based on velocity, pressure, temperature	EPA Reference Methods 1 and 2
Process Data	Plant instrumentation	--	--

VOC = volatile organic compound
THC = total hydrocarbon
FID = flame ionization detector
GC/MS = gas chromatography/mass spectrometry
CO = carbon monoxide
TCD = thermal conductivity detector

TABLE 5-2. SAMPLING AND ANALYTICAL PROCEDURES MATRIX - PLANT C-2[a]

Measurement Parameter	Sampling Procedure	Analytical Procedure	Reference
VOC Concentration (THC)	Rake-type probe, heated sampling line (145°C (293°F))	Inlet-modified FID-based gas gas chromatograph; outlet-FID-based THC analyzer	EPA Proposed Reference Method 25A
CO Concentration	Rake-type probe, heated sampling line (145°C (293°F))	FID-based analyzer with species separation and oxidation/reduction to methane	--
Molecular Weight (Fixed Gases)	Rake-type probe, heated sampling line (145°C (293°F))	TCD-based analyzer	EPA Reference Method 3 (Equivalent)
Velocity/Volumetric Flow Rate	S-type pitot tube	Calculations based on velocity, pressure, temper-	EPA Reference Methods 1 and 2
Process Data	Plant Instrumentation	--	--

[a]GC, GC/MS, and moisture samples were not taken because the catalyst was found to be deactivated (see Section 6.1).

TABLE 5-3. SAMPLING AND ANALYTICAL PROCEDURES MATRIX - PLANT C-3

Measurement Parameter	Sampling Procedure	Analytical Procedure	Reference
VOC Concentration (THC)	Rake-type probe, heated sampling line (145°C (293°F))	Inlet: modified FID-based gas chromatograph; outlet: FID-based THC analyzer	EPA Proposed Reference Method 25A
VOC Concentration (Speciated)	Rake-type probe, heated sampling line	FID-based gas chromatograph	EPA Draft Reference Method 18
VOC Species Identification	Rake-type probe, heated sampling line (145°C (293°F)); evacuated canisters for light organics, impinger condensate for heavy organics	GC/MS	--
CO Concentration	Rake-type probe, heated sampling (145°C (293°F))	FID-based analyzer with species separation and oxidation/reduction to methane	--
Molecular Weight (Fixed Gases)	Rake-type probe, heated sampling line (145°C (293°F))	TCD-based analyzer	EPA Reference Method 3 (Equivalent)
Moisture Content	Goose-neck probe, chilled impinger train	Gravimetric determination of condensate, measurement of sampled gas volume	EPA Reference Method 4
Velocity/Volumetric Flowrate	S-type pitot tube	Calculations based on velocity pressure, tempertuie	EPA Reference Methods 1 and 2
Process Data	Plant instrumentation	--	--

VOC = volatile organic compound
THC = total hydrocarbon
FID = flame ionization detector
GC/MS = gas chromatography/mass spectrometry
CO = carbon monoxide
TCD = thermal conductivity detector

TABLE 5-4. SAMPLING AND ANALYTICAL PROCEDURES MATRIX - PLANT C-4

Measurement Parameter	Sampling Procedure	Analytical Procedure	Reference
VOC Concentration (THC)	Rake-type probe, heated sampling line (100°C (212°F))	FID-based THC analyzers	EPA Proposed Reference Method 25A
VOC Concentration (Speciated)	Rake-type probe, heated sampling line (100°C (212°F))	FID-based gas chromato-	EPA Draft Reference Method 18
VOC Species Identification	Rake-type probe, heated sampling line (100°C, (212°F)), evacuated canisters	GC/MS	--
CO Concentration	Rake-type probe, heated sampling line (100°C (212°F))	FID-based analyzer with species separation and oxidation/reduction to methane	--
Molecular Weight (Fixed Gases)	Rake-type probe, heated sampling line (100°C (212°F))	TCD-based analyzer	EPA Reference Method 3 (Equivalent)
Moisture Content	Goose-neck probe, chilled impinger train	Gravemetric determination of condensate, measurement of sampled gas volume	EPA Reference Method 4
Velocity/Volumetric Flow Rate	S-type pitot tube	Calculations based on velocity, pressure, temperature	EPA Reference Methods 1 and 2
Process Data	Plant instrumentation	--	--

VOC = volatile organic compound
THC = total hydrocarbon
FID = flame ionization detector
GC/MS = gas chromatography/mass spectrometry
CO = carbon monoxide
TCD = thermal conducitivity detector

TABLE 5-5. SAMPLING AND ANALYTICAL PROCEDURES MATRIX - PLANT C-5

Measurement Parameter	Sampling Procedure	Analytical Procedure	Reference
VOC Concentration (THC)	Rake-type probe, heated sampling line (138°C (280°F))	Inlet: modified FID-based gas chromatograph; outlet: FID-based THC analyzer	EPA Proposed Reference Method 25A
VOC Concentration (Speciated)	Rake-type probe, heated sampling line (138°C (280°F))	FID-based gas chromatograph	EPA Draft Reference Method 18
VOC Species Identification	Rake-type probe, heated sampling line (138°C (280°F)); evacuated canisters for light organics, impinger, condensate for heavy organics	GC/MS	--
CO Concentration	Rake-type probe, heated sampling line (138°C (280°F))	FID-based analyzer with species separation and oxidation/reduction to methane	--
Molecular Weight (Fixed Gases)	O_2-gas sample drawn directly from outlet sampling port; CO_2 - rake-type probe, heated sampling line (138°C (280°F))	O_2-oxygen Fyrite apparatus; CO_2 - FID-based analyzer with species separation and oxidation/reduction to methane	EPA Reference Method 3 (Equivalent)
Moisture Content	Goose-neck probe, chilled impinger train	Gravimetric determination condensate, measurement of sampled gas volume	EPA Reference Method 4
Velocity/Volumetric Flowrate	S-type pitot tube	Calculations based on velocity, pressure, temperature	EPA Reference Methods 1 and 2
Process Data	Plant instrumentation	--	--

VOC = volatile organic compound
THC = total hydrocarbon
FID = flame ionization detector
GC/MS = gas chromatography/mass spectrometry
CO = carbon monoxide

TABLE 5-6. SAMPLING AND ANALYTICAL PROCEDURES MATRIX - PLANT C-6

Measurement Parameter	Sampling Procedure	Analytical Procedure	Reference
VOC Concentration (THC)	Rake-type probe, heated sampling line (100°C (212°F))	FID-based THC analyzers	EPA Proposed Reference Method 25A
VOC Concentration (Speciated)	Rake-type probe, heated sampling line (100°C (212°F))	FID-based gas chromatograph	EPA Draft Reference Method 18
VOC Species Identification	Rake-type probe, heated sampling line (100°C (212°F)), evacuated canisters	GC/MS	--
CO Concentration	Rake-type probe, heated sampling line (100°C (212°F))	FID-based analyzer with species separation and oxidation/reduction to methane	--
Molecular Weight (Fixed Gases)	O_2-gas sample drawn directly form outlet sampling port; CO_2 - rake-type probe heated sampling line (100°C (212°F))	O_2 - oxygen Fyrite CO_2 - FID-based analyzer with species separation and oxidation/reduction to methane	EPA Reference Method 3
Moisture Content	Goose-neck probe, chilled impinger train	Gravimetric determination of condensate, measurement of sampled gas volume.	EPA Reference Method 4
Velocity/Volumetric Flowrate	S-type pitot tube	Calculations based on velocity, pressure, temperature	EPA Reference Methods 1 and 2
Process Data	Plant instrumentation	--	--

VOC = volatile organic compound
THC = total hydrocarbon
FID = flame chromatography/mass spectrometry
CO = carbon monoxide
TCD = thermal conductivity detector

REFERENCES

1. EPA Proposed Method 25A, "Determination of Total Gaseous Organic Concentration Using a Flame Ionization Analyzer," Proposed Revision to Appendix A 40 CFR, Federal Register, Volume 45, Number 194, December 17, 1980. pp. 83149-83151.

2. EPA Draft Method 18, "Measurement of Gaseous Organic Compound Emissions by Gas Chromatography," Personal communication with Jeff Chappell, U.S. Environmental Protection Agency, IERL-RTP.

3. EPA Reference Method 3, "Gas Analysis for Carbon Dioxide, Oxygen, Excess Air, and Dry Molecular Weight," Appendix A to 40 CFR, Environmental Reporter, Bureau of National Affairs, Washington, D.C., December 5, 1980. pp. 33-36.

4. EPA Reference Method 4, "Determination of Moisture Content in Stack Gases," Appendix A to 40 CFR, Environmental Reporter, Bureau of National Affairs, Washington, D.C., December 5, 1980. pp. 36-41.

5. EPA Reference Method 1, "Sample and Velocity Traverses for Stationary Sources." Appendix A to 40 CFR, Environment Reporter. Bureau of National Affairs. Washington, D.C. December 5, 1980. pp. 23-25.

6. EPA Reference Method 2, "Determination of Stack Gas Velocity and Volumetric Flow Rate (Type S Pitot Tube), Appendix A to 40 CFR, Environmental Reporter, Bureau of National Affairs, Washington, D.C., December 5, 1980. pp. 25-33.

6. Results of Testing and Analysis

Catalytic incinerator performance data were collected on eight different incinerators at six industrial sites. These incinerators were applied in the can coating, coil coating, magnet wire and graphic arts printing industries.

Detailed discussions of the test results for each site are presented in the Site Test Reports, Volumes 2-8 of this report. The purpose of this section is to summarize and compare the results from all six sites. No attempt is made in this section to repeat all of the detailed discussions in the Site Test Reports. However, information considered necessary to provide an objective comparison of the performance of the eight incinerators is included.

Waste gas characteristics and incinerator operating conditions are discussed first. This information is then used as the basis for a comparative analysis of incinerator performance during typical plant and incinerator operation. A comparison of the effect of varying major operating parameters at each site is presented last.

6.1 WASTE GAS CHARACTERISTICS AND INCINERATOR OPERATING CONDITIONS

Waste gas characteristics and incinerator design and operating parameters for the eight incinerators are presented in Table 6-1. Each of the different characteristics and parameters shown in Table 6-1 are discussed in this section. Test methods used to obtain the different parameters are discussed in Section 5.0 and in the Site Test Reports.

TABLE 6-1. WASTE GAS CHARACTERISTICS AND INCINERATOR DESIGN AND OPERATING PARAMETERS

Plant ID (Incinerator No.)	Process Type	Waste Gas Flowrate[a] Nm^3/s (scfm)	Waste Gas Temperature °C (°F)	Major Solvents Identified[b]	Average Temperatures[c] Catalyst Inlet/Outlet °C °F	Catalyst Bed Pressure Drop kPa ("H_2O)	Catalyst Type	Catalyst Age (months)	SLA Residence Time (sec)	Design Destruction Efficiency (%)
Plant C-1 (27)	Can Coating	3.6 (7600)[d]	121 (250)	toluene, xylenes, ethyl benzenes	363/396 (685/745)	0.60 (2.4)	Ceramic honeycomb	8	0.03	90
Plant C-1 (28)	Can Coating	3.2 (6780)[d]	89 (190)	MIBK,[e] xylenes[e], methyl ethyl benzenes	316/410 (600/770)	0.57 (2.3)	Ceramic honeycomb	23/<1[f]	0.03	90
Plant C-4	Can Coating	2.52 (5330)	164 (327)	MIBK, cellosolve, xylenes, ethylbenzene[e]	332/393 (630/740)	0.42 (1.7)	Ceramic honeycomb	11[g]	0.03	90
Plant C-2	Graphic Arts Printing	0.9 (2000)	127 (260)	C12 to C18 hydrocarbons	493/438 (920/820)[h]	2.5 (10)	Ceramic honeycomb	i	NA	90
Plant C-3 (20)	Graphic Arts Printing	1.26 (2660)	177 (350)	C12 to C18 hydrocarbons	372/379 (701/713)	0.25 (1.0)	Ceramic spherical pellets	16	0.04[j]	90
Plant C-3 (21)	Graphic Arts Printing	2.21 (4670)	178 (352)	C12 to C18 hydrocarbons	353/412 (667/774)	0.25 (1.0)	Ceramic spherical pellets	8	0.04[k]	90
Plant C-5	Magnet Wire	0.28 (603)	233 (451)	phenol, cresols	393/507 (740/945)	0.17 (0.7)	Metal ribbons	<1	0.015[k]	NA[l]
Plant C-6	Coil Coating	5.33(11300)	143 (290)	MEK, toluene[e]	285/427 (545/800)	0.42 (1.7)	Ceramic honeycomb	1	0.03	95

FOOTNOTES FOR TABLE 6-1

[a]This is the total inlet gas flow to the incinerator on a wet basis.

[b]Solvents were identified by gas chromatography and/or mass spectroscopy.

[c]These temperatures represent the average temperatures measured for each incinerator during typical incinerator operation. Tests conducted at different operating temperatures are discussed in Section 6.3.

[d]These flowrates were calculated from a CO_2 mass balance around SLA fan (See Appendix A, Plant C-1 Site Test Report). Flows were not measured because no acceptable sampling location was available.

[e]Solvent formulations varied for different coatings. MIBK = methyl isobutyl ketone, MEK = methyl ethyl ketone.

[f]Two catalyst ages were tested at this plant.

[g]Catalyst was cleaned 5 months prior to test.

[h]This catalyst is believed to have been deactivated by either high temperature operation or material masking. The temperature drop across the catalyst bed and a 2.5 kPa (10"H_2O) pressure drop measured for this catalyst is evidence of the deactivation.

[i]The catalyst in the incinerator was 8 years old but the operating hours logged by the incinerator is much less than 8 years. The actual number of hours logged is, however, unknown.

[j]Assumes 40 percent void fraction.

[k]Assumes 50 percent void fraction.

[l]Design efficiency unknown because incinerator was modified by plant to increase catalyst volume.

Waste gas flowrates ranged widely for the eight incinerators from 0.28 to 5.33 Nm^3/s (600 to 11,300 scfm). The smallest flowrate was for the magnet wire plant while the largest was for the coil coating plant.

Comparison of waste gas temperatures shows a fairly narrow range from 89 to 177°C (190 to 350°F) for all plants except for the magnet wire plant. The magnet wire coater has a relatively high waste gas temperature because the curing oven in this process operates at a higher temperature.

The types of solvents destroyed in the catalytic incinerators varied considerably for the different industries. The inlet concentration of the solvents ranged from approximately 5 to 15 percent of the lower explosive limit (LEL). The can coating solvents generally consisted of aromatic compounds, such as toluene, xylenes and ethyl benzenes. Some of these coatings also contained methyl isobutyl ketone (MIBK) and/or cellosolve.

The inks used at the graphic arts printing plants contained a heavy organic solvent consisting of C_{12} through C_{18} hydrocarbons with an average molecular weight of about 210 g/mole.

The solvents in the magnet wire coatings were phenols and cresols and the coil coating solvents were methyl ethyl ketone (MEK) and toluene.

Average catalyst bed inlet and outlet temperatures were generally similar at the different plants except for Plants C-2 and C-5. At Plant C-2, the catalyst outlet temperature consistently averaged about 56°C (100°F) lower than the catalyst inlet temperature. This temperature drop suggests that no solvent destruction was occurring on the catalyst bed. These temperature readings were confirmed by multiple readings with several thermocouples and readout devices. The temperature drop across the catalyst bed, in conjunction with the very high catalyst bed pressure drop, is evidence that the Plant C-2 catalyst was fouled or deactivated and unable to promote solvent oxidation.[1] The catalyst was probably damaged by either high temperature operation (7540°C, 1000°F) or material masking.

At the magnet wire plant, Plant C-5, the incinerator is typically operated at catalyst inlet and outlet temperatures higher than at the other plants. This higher operating temperature is necessary to avoid the possible condensation of coating resin on the catalyst bed. A small amount of the coating resin is carried out of the curing and drying oven with the waste gas.

Comparison of the catalyst types and residence times for the eight incinerators shows that the lowest SLA residence time is for the metal ribbon catalyst at Plant C-5. As discussed in Section 4.5, this incinerator was originally installed circa 1967 and was modified by the plant in 1974 to increase the catalyst volume. However, even with the increased catalyst volume, the incinerator still has a SLA residence time that is lower than the other more recently installed incinerators tested.

The age or service life of the catalyst beds tested ranged from several days to 23 months at all plants except C-2. At Plant C-2, the catalyst bed was 8 years old but initial plant information indicated that the incinerator had logged only 2000 hours of service since startup. This estimate was later found to be considerably in error but the actual service life of the catalyst is still unknown.

At Plants C-1 and C-5, the catalyst was replaced during the test period. The first catalyst tested at Plant C-5 was also of unknown age. Data for this catalyst is not presented in this report but is discussed in the Site Test Report for Plant C-5.

The design destruction efficiency for six of the eight incinerators is 90 percent and one incinerator has a 95 percent design efficiency. The design destruction efficiency for Plant C-5 is not available.

6.2 RESULTS OF COMPARATIVE ANALYSIS

A comparative analysis of performance for the eight catalytic incinerators is presented in this section. This analysis is based on the waste gas characteristics and incinerator operating conditions discussed in Section 6.1. The performance of the incinerators is compared in terms of solvent destruction efficiency, outlet gas solvent concentrations and energy usage.

Incinerator inlet and outlet concentrations presented in this section are based on EPA Method 25A analyses. The concentrations are reported as parts per million by volume carbon (ppmv) but were quantitated against propane standards. Destruction efficiencies presented in this section were calculated on the basis of inlet and outlet solvent mass flowrates except for Plant C-1. At Plant C-1, inlet and outlet gas flowrates could not be measured according to EPA Method 2 and these efficiencies were calculated directly from measured inlet and outlet Method 25A concentrations. Examples of the efficiency calculations are presented in Appendix A.

Method 25A concentrations, destruction efficiencies and energy usages for the eight catalytic incinerators are presented in Table 6-2. Two data sets are presented for incinerator 28 at Plant C-1. Each of the data sets for incinerator 28 represents a different catalyst age. The data in Table 6-2 were collected under steady state process and incinerator operating conditions.

6.2.1 Destruction Efficiency

As shown in Table 6-2, destruction efficiencies for the eight incinerators varied widely, from 80.1 to 98.6 percent. Major factors causing this variation include catalyst age or condition and SLA residence time.

TABLE 6-2. SUMMARY OF PERFORMANCE DATA FOR TYPICAL OPERATING CONDITIONS[a]

Plant ID (Incinerator No.)[b]	Type of Process	Coatings Testing	Hours of Data[c]	Method 25A Concentrations[d] Inlet (ppmv)	Outlet (ppmv)	Destruction Efficiency (%)[e]	Average Energy Usage[f] kJ/Nm^3	(Btu/scf)	Heat Recovery
Plant C-1 (27)	Can Coating	1	27.6	4000-5810	181-275	95.4-96.4	270	(7.2)	39[j]
Plant C-1 (28)	Can Coating	3	7.8	2840-7760	173-321	93.4-95.9	230	(6.1)	46[j]
Plant C-1 (28)	Can Coating	7	26	2270-5755	46-341	96.3-98.6 (91.7-93.4)[g]	230	(6.1)	46[j]
Plant C-4	Can Coating	2	13	5480-7560	385-687	88.7/94.0	260	(6.9)	None
Plant C-2	Graphic Arts Printing	1	23	1020	169	81.2	780	(21)	None
Plant C-3 (20)	Graphic Arts Printing	1	5.5	1240	241	80.1	160	(4.3)[i]	36%[k]
Plant C-3 (21)	Graphic Arts Printing	5	7.2	1370-4030	90-165	93.4-95.9	80	(2.1)[i]	66%[k]
Plant C-5	Magnet Wire	1	13.8	8720	1590	80.6	240	(6.4)[i]	None
Plant C-6	Coil Coating	5	27	6220-12,860[h]	272-305	96.5-97.5	150	(3.9)	31%[j]

FOOTNOTES FOR TABLE 6-2

[a]Data presented in this table are for steady state incinerator and coating process operating conditions.

[b]Tests were conducted between November 1982 and March 1983.

[c]Represents hours of data collected at "typical" incinerator operating temperature (see Table 6-1). Data collected at other operating temperatures are discussed in Section 6.3.

[d]Concentrations are for the incinerator inlet/outlet gas streams. The range of values represents the range of average concentrations measured during monitoring periods for the different coatings tested. Monitoring period lengths generally exceeded 1 hour and ranged up to about 14 hours. Method 25A results for different monitoring periods are summarized in the Site Test Reports. ppmv = parts per million by volume carbon, quanitated against propane standards.

[e]For all plants except Plant C-1,

$$\text{Destruction Efficiency} = \frac{Min - Mout}{Min} \times 100\%, \text{ where}$$

$$M = \text{VOC mass flowrate} = \frac{V \text{ moles C}}{10^6 \text{ moles gas}} \times \frac{12g \text{ C}}{\text{mole C}} \times \frac{\text{moles gas}}{0.024 \text{ m}^3} \times \frac{Q \text{ m}^3 \text{ gas}}{\text{sec}}$$

V = VOC concentration, ppmv (in or out)

Q = gas flowrate, Nm^3/sec (in or out)

For plant C-1, $\text{Destruction Efficiency} = \frac{Vin - Vout}{Vin} \times 100\%$.

[f]Energy usage is for fuel gas required to heat waste gas to catalyst inlet temperature.

[g]These values are for one of the 7 coatings tested. This solvent formulation apparently had a greater percentage of MIBK and a higher average solvent molecular weight than the other formulations tested.

[h]This concentration is for "main inlet" gas stream as shown in Figure 4-6.

[i]Energy usage calculated based on waste gas temperature and catalyst bed inlet temperature.

[j]Heat recovered by recycling incinerator outlet gas to drying oven. Represents percent recovery of heat generated by natural gas and fuel combustion in the incinerator.

[k]Heat recovered by recuperative heat exchange. Represents percent recovery of heat required to raise the waste gas temperature to the catalyst bed inlet temperature.

Of the eight incinerators, four achieved destruction efficiencies in excess of 90 percent on all of the coatings tested. Two of these incinerators achieved efficiencies in excess of 95 percent for all coatings. Three of the four incinerators showing efficiencies greater than 90 percent used ceramic-based honeycomb catalyst, with a 0.03 second residence time and one used ceramic-based spherical pellet catalyst, with a 0.04 second residence time. The catalyst ages in these incinerators ranged from new to 23 months.

The Plant C-4 incinerator showed destruction efficiencies of 88.4 and 94 percent, respectively for the two coatings tested. These efficiencies are slightly lower on average than those measured for the similarly designed incinerators at Plant C-1. As noted in Section 4.4, the catalyst in the Plant C-4 incinerator had become masked during the first 6 months of operation and was cleaned with an alkaline solution 5 months prior to testing. One possible explanation for the lower efficiencies measured at Plant C-4 compared to Plant C-1 is that the catalyst was still partially masked.

Destruction efficiencies of approximately 80 percent were observed for three of the eight incinerators: Plants C-2, C-3 (20) and C-5. The lower efficiencies for these systems can be attributed to either the catalyst condition or the catalyst bed residence time.

For Plant C-2, the comparatively low destruction efficiency of 81 percent is probably due to the catalyst being fouled. As discussed in Section 6.1, a temperature drop and a high pressure drop were measured for this catalyst. These conditions suggest that the catalyst was fouled or deactivated and unable to promote solvent oxidation[1]. The solvent destruction observed for the incinerator must have occurred in the burner zone of the incinerator instead of in the catalyst bed.

For Plant C-5, the comparatively low efficiency of 81 percent is most likely due to the shorter SLA residence time. The residence time for this incinerator was 0.015 seconds compared with 0.03 seconds or greater for the incinerators showing destruction efficiencies in excess of 90 percent. As discussed in Section 6.1, the Plant C-5 incinerator was originally installed circa 1967 and was modified by the plant in 1974 to increase the catalyst volume. The current catalyst volume still provides only half the residence time of newly installed systems.

For Plant C-3, the only plausible explanation for the comparatively low efficiency is normal catalyst aging. No data are available that suggest any particular problems that may have resulted in deactivation of this catalyst. Comparison of incinerators 20 and 21 at Plant C-3 shows that the major difference between the incinerators is the service life or age of the catalysts, at 16 and 8 months, respectively.

The 16 month-old catalyst in incinerator 20 also shows a much lower destruction efficiency than the 23 month-old catalyst tested at Plant C-1 (28). This is somewhat unexpected because catalyst activity generally decreases with time. However, the catalyst geometries and solvent types differ for the two incinerators and the conditions effecting catalyst deactivation are not necessarily the same at both plants.

Comparison of the measured destruction efficiencies presented in Table 6-2 with the design efficiencies presented in Table 6-1 shows that two incinerators operated considerably below design levels. These were the Plant C-2 incinerator and incinerator 20 at Plant C-3. As discussed above, the lower efficiency at Plant C-2 was due to the catalyst being fouled and the lower efficiency for incinerator 20 was probably due to catalyst age.

A comparison of Method 25A and Method 18 THC destruction efficiencies is shown in Table 6-3. Data for different incinerator operating temperatures and different products are shown in Table 6-3 for Plants C-4, C-5, and C-6. The THC destruction efficiencies measured according to the two test methods are in good agreement.

TABLE 6-3. COMPARISON OF METHOD 25A AND METHOD 18 THC DESTRUCTION EFFICIENCIES

Plant ID	Product	Catalyst Temperature Inlet/Outlet °C	(°F)	Method 18 THC Destruction Efficiency	Method 25A Destruction Efficiency
C-4	A	318/382	(605/720)	95.6	94.0
C-4	B	335/397	(635/746)	89.4	88.7
C-4	B	368/432	(695/810)	92.3	91.0
C-4	B	391/452	(735/845)	93.6	92.4
C-5	A	402/482	(755/900)	78.9	79.5
C-5	A	468/554	(875/1030)	84.4	82.4
C-5	A	493/577	(920/1070)	83.3	82.0
C-6	A	285/443	(545/830)	96.3	97.5
C-6	B	341/443	(645/830)	96.6	97.0
C-6	D	288/374	(550/705)	94.3	95.9
C-6	E	277/435	(530/815)	96.1	97.3

6.2.2 Outlet Method 25A Concentrations

Outlet Method 25A concentrations measured for the eight incinerators varied fairly widely. As shown in Table 6-2, the lowest concentration of 46 ppmv was for Plant C-1, incinerator 28 and the highest, 1600 ppmv, was for Plant C-5.

All of the incinerators showed outlet concentrations less than 700 ppmv except for Plant C-5.

6.2.3 Energy Usage

Average energy usage for the eight incinerators in terms of energy per unit volume of waste gas varied over a fairly narrow range except for Plant C-2. For plants other than C-2, energy usage ranged from 78 to 270 kJ/m^3 (2.1 to 7.2 Btu/scf). At Plant C-2, the catalyst bed inlet temperature was 110 to 170°C (200 to 300°F) higher than at the other plants and energy usage was estimated to be 780 kJ/m^3 (21 Btu/scf).

Five of the eight incinerators employed some type of heat recovery. At Plants C-1 and C-6, outlet gas from the incinerator was recycled to the drying oven and at Plant C-3 recuperative heat exchangers were used.

For the plants using gas recycle, heat recovery rates ranged from 31 to 46 percent. This heat recovery represents the percentage of heat generated by natural gas and solvent combustion that is reused in the process (see Appendix B).

At Plant C-3, heat recovery was determined to be 36 and 66 percent for incinerators 20 and 21, respectively. This heat recovery represents the percent recovery of heat required to raise the waste gas temperature to the catalyst bed inlet temperature.

6.3 EFFECT OF VARYING OPERATING PARAMETERS

Parametric analyses were performed on the Method 25A results for each incinerator to assess the effect of operating parameter variations on performance. Incinerator operating temperature and inlet solvent concentrations were found to be important parameters. The effect of these parameters on incinerator performance is discussed in detail in the Site Test Reports and is compared for the different incinerators in this section.

The Method 18 results for each incinerator were analyzed in an attempt to determine the effect of solvent type on performance. These results appear to show weak trends in the relative destruction efficiencies of different solvents. However, few conclusions could be drawn from these results because of large statistical variability in the data. Confidence intervals for the average destruction efficiencies measured for different solvents overlap because the number of samples taken is statistically small and the standard deviation of the measured concentrations is large. Analyses of the Method 18 data for each plant are included in the Site Test Reports but are not repeated in this section.

Data presented in this section were taken directly from the Site Test Reports except where otherwise noted. No data are presented in this section for Plant C-2 because operating conditions were not varied at this plant. Similarly, Plant C-1 data are not included in the temperature analysis because the operating temperature was not varied.

Catalyst Operating Temperature

At Plants C-3 through C-6, the incinerator operating temperature was varied intentionally to observe the effect on incinerator performance. For purpose of this analysis, the catalyst inlet temperature was taken to represent the catalyst operating temperature. The effect of catalyst inlet temperature on VOC destruction efficiency for the five incinerators tested at Plants C-3 through C-6 is illustrated in Figure 6-1. Figure 6-2 shows the effect of catalyst inlet temperature on outlet THC concentration of the five incinerators. All of the data presented in Figures 6-1 and 6-2 represent steady-state monitoring periods at a relatively constant inlet THC concentration for each of the incinerators tested.

As shown in Figure 6-1, increasing the catalyst operating temperature resulted in increased destruction efficiency. For two of the five incinerators, the effect of catalyst temperature on destruction efficiency was relatively small. The incinerators at Plants C-3 and C-5 showed a fairly large effect of temperature on efficiency while Plants C-4 and C-6 showed a relatively small effect. The greatest effect of temperature was observed for incinerator 20 at Plant C-3, where a 25°F increase in catalyst inlet temperature resulted in a 2.5 percent increase in destruction efficiency.

Figure 6-2 demonstrates that, as expected, increasing the catalyst operating temperature resulted in decreased outlet Method 25A concentrations for the five incinerators shown. As with destruction efficiency, the outlet Method 25A concentrations were most greatly affected by changes in catalyst operating temperature at Plants C-3 and C-4. The effect of catalyst operating temperature on outlet Method 25A concentration was smaller at Plants C-5 and C-6.

Inlet Method 25A Concentration

The THC concentration of the inlet SLA varied during testing at each of the incinerators due to changes in product type, product size, solvents used, and line speeds. The effect of variations in inlet Method 25A concentration is shown in Figure 6-3 for seven incinerators. Inlet concentrations shown in Figure 6-3 for Plant C-6 were calculated from inlet THC mass flowrates presented in the Interim Test Report for this plant. An example of the calculation is shown in Appendix C.

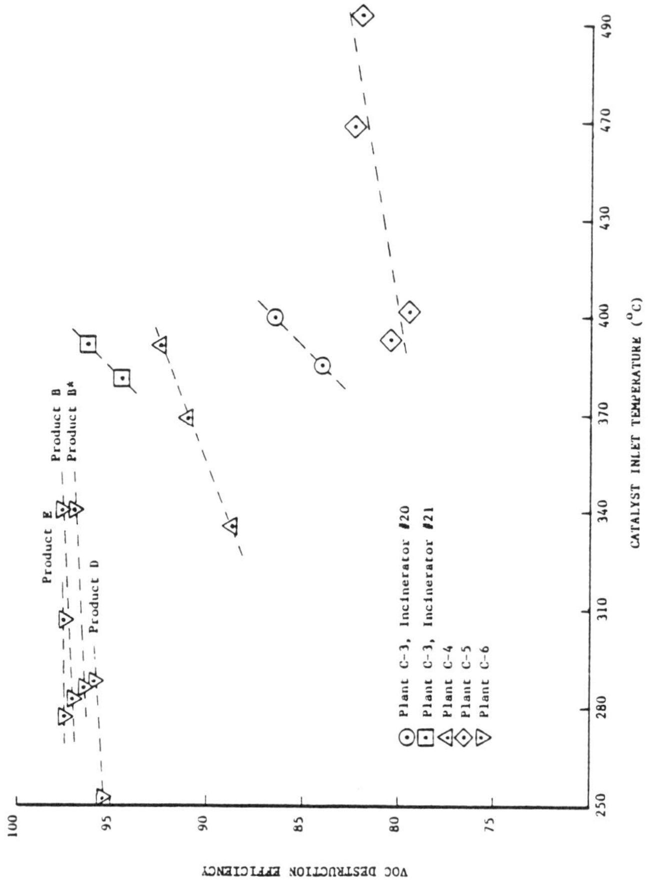

Figure 6-1. Effect of catalyst inlet temperature on destruction efficiency.

*Data is presented for two different inlet concentrations of Product B.

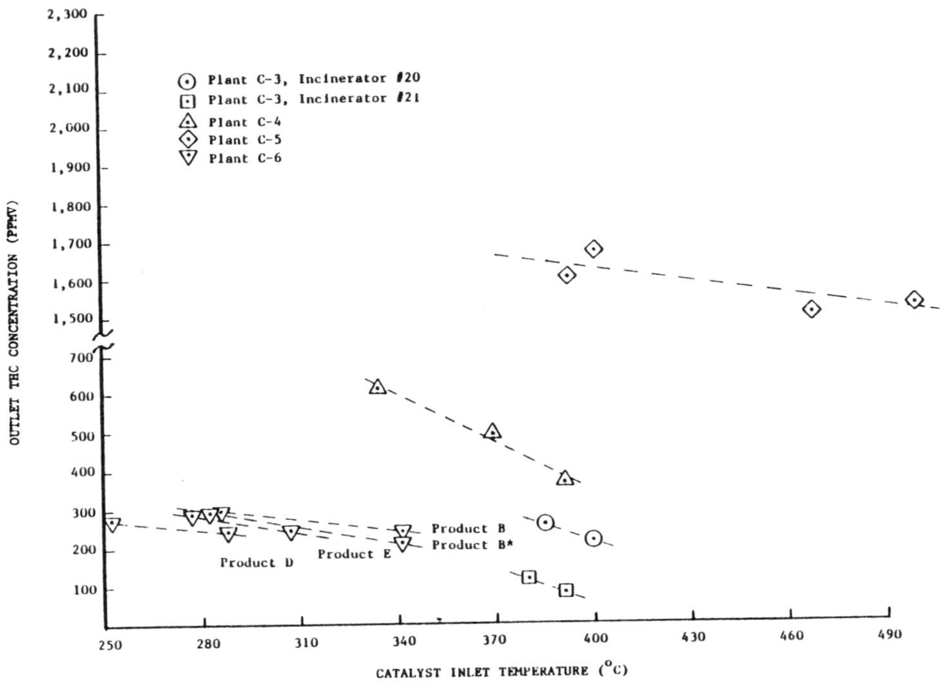

Figure 6-2. Effect of catalyst inlet temperature on outlet Method 25A concentration.

*Data is presented for two different inlet concentrations of Product B.

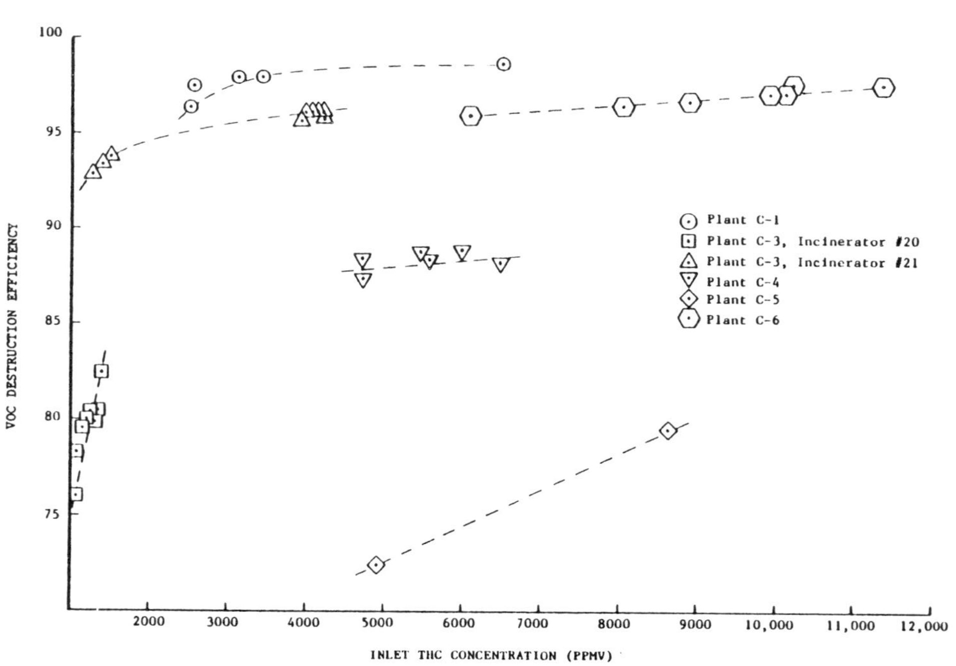

Figure 6-3. Effect of inlet Method 25A concentration on destruction efficiency.

*For Plant C-6, inlet THC mass flowrates presented in the interim test report were converted to "equivalent" concentration (see Appendix C) for preparation of this figure.

As shown in Figure 6-3, destruction efficiency increased in all cases with increased inlet Method 25A concentration. The effect of increased inlet Method 25A concentration appeared to be greatest at lower destruction efficiencies. A change in inlet Method 25A concentration had the greatest effect on destruction efficiency for incinerator 20 at Plant C-3 and the incinerator at Plant C-5. Both of these incinerators operated at destruction efficiencies less than 85 percent. The effect of inlet Method 25A concentration on destruction efficiency appears to be smaller for the remaining incinerators, all of which showed higher destruction efficiencies.

REFERENCES

1. Telecon. E. Smith, Johnson Matthey, Inc., with J. R. Blacksmith, Radian Corporation. Conversation about Plant C-2 incinerator. February 8, 1983.

7. Recommendations

The EPA catalytic incinerator program was initiated to assemble analyze, assess and develop information on the use and the ability to use catalytic incinerators for VOC control. The program concentrated on the use of the technology for control of VOC process emissions. Section 1 details the two primary objectives for the total program and states (objective 2) the objective of the current testing program: to gather actual performance data on operating industrial catalytic incinerators. The objective of the testing program has been successfully met for four major industrial users of catalytic incinerators including: can coating, coil coating, web off-set printing and magnet wire coating.

Work from several catalytic incinerator projects has vastly increased the information base from which both EPA and industrial users can draw useful conclusions. However, there are still data gaps and information deficiencies that when filled, will make the total technology assessment of catalytic incineration more complete. Recommendations for future work in catalytic incineration will compromise the remainder of this section.

Future work in the area of catalytic incineration should be focussed on providing potential industry users and the EPA with additional information needed to assess the applicability and potential performance of incinerators applied in industries other than those tested in this program. Information that will be required to meet this objective includes (1) a better understanding of the effect of solvent type on catalytic incinerator performance and (2) performance data on solvent types significantly different than those tested in this program.

Further data on the effect of solvent type on incinerator performance should be collected. This data will allow the prediction of potential destruction efficiencies achievable on processes using similar or more easily destroyed solvents than those tested. This data should include actual destruction data but should also consider the chemical/physical makeup of the solvents. In this way general trends might be established between the solvents for which there is destruction data and for solvents for which there is no data, but there is chemical/physical similarity. This type of information would be very useful in a predictive way for anyone desiring to estimate the destruction efficiency for a given solvent.

REFERENCES

1. U.S. Environmental Protection Agency. Control of Volatile Organic Emissions from Existing Sources. Volume 8: Graphic Arts - Rotogravure and Flexography. Office of Air Quality Planning and Standards Pb No. EPA-450/2-78-033. December 1977.

Appendix A

Example Calculations

Methods used to calculate destruction efficiency and heat recovery are presented in the following subsections.

A.1 DESTRUCTION EFFICIENCY

For all plants except Plant C-1, the destruction efficiency is based on inlet and outlet THC mass rates as follows:

$$DE = \frac{Min - Mout}{Min} \times 100\%$$

where DE = destruction efficiency, %

Min = inlet THC mass rate, g C/sec

Mout = outlet THC mass rate, g C/sec

THC mass rates are based on measured Method 25A concentrations and gas flowrates:

$$M = \frac{(V)\ \text{moles C}}{10^6\ \text{moles gas}} \times \frac{12\text{gC}}{\text{mole C.}} \times \frac{\text{mole gas}}{0.024\ m^3\ \text{gas}} \times \frac{(Q)\ m^3\ \text{gas}}{\text{sec}}$$

where M = THC mass rate (inlet or outlet) g C/sec

V = THC concentration (inlet or outlet), ppmv

Q = gas flowrate (inlet or outlet), Nm^3/sec

Standard or normal conditions refer to one atmosphere of pressure and 20°C (68°F).

For Plant C-1, inlet and outlet gas flowrates could not be measured because no acceptable location was available. Destruction efficiency for this plant was based on Method 25A inlet and outlet concentrations as follows:

$$DE = \frac{Vin - Vout}{Vin}$$

where V = THC concentration (inlet or outlet), ppmv

A.2 HEAT RECOVERY BY RECUPERATIVE HEAT EXCHANGE

For recuperative exchangers, the heat recovery is defined as the percentage of heat provided by the heat recovery system of the total required to raise the inlet solvent laden air (SLA) temperature to the catalyst inlet temperature:

$$HR = \frac{HEX}{Q_{TOTAL}} \times 100\%$$

where H = heat recovery, %

Q_{HER} = heat added to the inlet SLA by heat exchange with the incinerator exhaust gas, J/sec

Q_{TOTAL} = heat necessary to raise the inlet SLA temperature to the catalyst inlet temperature, J/sec

The above equation is solved by making the following assumptions and substitutions:

$$HR = \frac{mCp \ T_{SLA,HEXout} - T_{SLA,HEXin}}{mCp \ T_{SLA,catin} - T_{SLA,HEXin}} \times 100\%$$

where m = mass flowrate of SLA, gmole/sec

Cp = specific heat of SLA at the elevated temperatures, J/gmole°K

$T_{SLA,HEXout}$ = temperature of the SLA exiting the heat exchanger, °K

$T_{SLA,HEXin}$ = temperatures of the SLA entering the heat exchanger, °K

$T_{SLA,catin}$ = temperature of the SLA entering the catalyst bed, °K

Of the above temperatures, $T_{SLA,HEXout}$ was not measured. It was calculated by performing an energy balance around the heat exchanger:

$$mCp\,(T_{SLA,HEXout} - T_{SLA,HEXin}) = mCp\,(T_{exh,HEXin} - T_{exh,HEXout})$$

where $T_{exh,HEXin}$ = temperature of the exhaust gas entering the heat exchanger, °K

$T_{exh,HEXout}$ = temperature of the exhaust gas exiting the heat exchanger, °K

The SLA and exhaust gas mass rates are assumed to be equal for this calculation, since the contribution by the fuel gas to the exhaust gas mass rate is negligible. Specific heats of the inlet and outlet streams are also assumed to be equal over the measured temperature range.

A.3 HEAT RECOVERY BY EXHAUST GAS RECYCLE

For systems with outlet gas recycle, heat recovery was defined as the percent of heat generated by natural gas and solvent combustion that is reused in the process. The heat reused in the process was assumed to be equivalent to all heat not exiting the process with the stack exhaust gas. Heat recovery was then calculated as follows:

$$HR = \left[1 - \frac{Q_{ex}}{Q_{gen}}\right] \times 100\%$$

where HR = heat recovery, %

Q_{ex} = heat added to stack exhaust gas by natural gas

and solvent combustion, J/sec

Q_{gen} = total heat generated by natural gas and solvent combustion, J/sec

The above equation is solved by making the following assumptions and and substitutions:

$$HR = [1 - \frac{m_{stack}\ Cp(T_{stack} - T_{sys\ in})}{m_{SLA}\ Cp(T_{catout} - T_{SLA\ in})}] \times 100\%$$

where m = gas mass flowrate (stack gas or SLA), g mole/sec

Cp = average specific heat of SLA for the range of temperature, J/g mole °K

T_{stack} = temperature of the stack exhaust gas, °K

$T_{sys\ in}$ = temperature of system in leakage air entering the ends of the drying oven, °K

T_{catout} = temperature of SLA exiting the catalyst bed, °K

$T_{SLA\ in}$ = temperature of SLA entering the incinerator, °K

The temperature of the in leakage air is equal to the ambient temperature of the coating room near the oven or approximately 32°C (90°F).

Appendix B

Summaries of Method 25A Results

The results of Method 25A analyses for the eight incinerators tested at Plants C-1 through C-6 are presented in Tables B-1 through B-8. Results presented in the tables represent all data collected under normal process operation at each of the plants. Periods during which process operation was unsteady are noted. Different solvents used during testing at each incinerator are identified by letter.

TABLE B-1. SUMMARY OF METHOD 25A RESULTS FOR INCINERATOR 27 - PLANT C-1

Date	Monitoring Period	Length of Monitoring Period (min)	Product ID	Solvents Added to Coating***	Inlet Method 25A Concentration (ppmv)*	Outlet Method 25A Concentration (ppmv)*	Average Destruction Efficiency (%)**	Average Gas Temperature °C (°F) Catalyst Inlet	Average Gas Temperature °C (°F) Catalyst Outlet	Outlet CO (ppmv)
11/16/82	13:03-14:04	62	F	mineral spirits	1366†	††	NA	388 (730)	393 (740)	—
11/16/82	14:05-16:55	171	F	mineral spirits	1767	††	NA	388 (730)	393 (740)	—
11/16/82-11/17/82	20:44-03:22	399	F	mineral spirits	2220	††	NA	374 (705)	407 (765)	—
11/17/82	05:14-09:07	234	G	50% mineral spirits 30% isophorone 20% DIBK	3356	††	NA	374 (705)	404 (760)	—
11/19/82	20:04-20:21	18	K	butyl cellosolve	1796†	197	89.0†	388 (730)	393 (740)	
11/19/82-11/20/82	22:35-05:52	438	K	butyl cellosolve	4069	181	95.6	357 (675)	396 (745)	—
11/22/82	07:12-98:23	72	K	butyl cellosolve	2158†††	235	89.1†††	354 (670)	399 (750)	—
11/27/82	09:32-11:31	120	K	butyl cellosolve	5827	212	96.4	338 (640)	399 (750)	13
11/22/82	12:22-19:48	447	K	butyl cellosolve	4460	205	95.4	343 (650)	393 (740)	—
11/22/82-11/23/82	20:50-06:01	552	K	butyl cellosolve	4964	225	95.5	343 (650)	393 (740)	—
11/23/82	06:09-07:42	96	K	butyl cellosolve	5582	275	95.1	341 (645)	393 (740)	10

*ppmv = parts per million as C_1, quantitated against propane.

**Destruction Efficiency $= \frac{V_{in} - V_{out}}{V_{in}} \times 100\%$ where, V = THC concentration (in or out), ppmv.

***These solvents are added in small amounts to premixed coatings purchased by Plant C-1. DIBK = diisobutyl ketone, MIBK = methylisobutyl ketone.

†Process problems — unsteady operation.

††These data have been invalidated due to a suspected leak in the sample acquisition system. THC values measured on the Byron 401 were less than the inlet gas CH_4 concentrations.

†††These data were collected during process startup.

TABLE B-2. SUMMARY OF METHOD 25A SAMPLING RESULTS FOR INCINERATOR 28 - PLANT C-1

Date	Monitoring Period	Length of Monitoring Period (min)	Product ID	Solvents Added to Coating***	Inlet Method 25A Concentration (ppmv)*	Outlet Method 25A Concentration (ppmv)*	Average Destruction Efficiency (%)**	Average Gas Temperature °C (°F) Catalyst Inlet	Average Gas Temperature °C (°F) Catalyst Outlet	Outlet CO (ppmv)
11/12/82	20:03-21:09	24†	A	butyl cellosolve	4053††	201	93.0	202 (540)	402 (755)	–
11/12/82-11/13/82	22:31-0.2:33	243	B	50% mineral spirits 50% butyl cellosolve	2848	173	93.9	313 (595)	399 (750)	–
11/13/82	03:22-06:42	201	C	80% mineral spirits 20% isophorone	7809	321	95.9	296 (565)	457 (855)	–
11/15/82	21:10-21:42	33	D	mineral spirits	2400	89	96.3	346 (655)	396 (745)	–
11/15/82	22:58-23:18	21	D	mineral spirits	685†††	51	92.6†††	332 (630)	404 (760)	–
11/16/82	00:48-02:05	78	D	mineral spirits	777†††	44	94.3†††	327 (620)	399 (750)	–
11/16/82	02:06-03:35	90	D	mineral spirits	2803	95	96.6	329 (625)	396 (745)	–
11/16/82	04:39-06:57	141	K	mineral spirits	2293	65	97.2	332 (630)	396 (745)	–
11/19/82	00:57-03:50	174	I	DIBK***	2317	58	97.5	344 (651)	399 (750)	–
11/19/82	04:21-08:08	228	J	mineral spirits	3161	68	97.8	324 (615)	399 (750)	–
11/19/82	09:24-10:59	96	J	mineral spirits	3095	68	97.8	321 (610)	398 (750)	3
11/19/82	11:31-12:30	60	J	mineral spirits	1498†	34	97.7†	349 (660)	396(745)	1
11/19/82	14:22-15:24	63	A	butyl cellosolve	5762	84	98.5	308 (585)	407 (760)	–
11/19/82	15:51-16:38	48	A	butyl cellosolve	3409	46	98.7	349 (660)	402 (755)	–
11/23/82	18:24-20:01	98	L	MIBK***	4142§	341	91.8	266 (510)	452 (845)	–
11/23/82	20:22-23:18	177	L	MIBK	2972§	186	93.7	279 (535)	436 (820)	–
11/24/82	00:47-07:35	409	L	MIBK	4124§	274	93.4	285 (545)	429 (805)	–

*ppmv = parts per million as C_1, quantitated against propane. NMHC = nonmethane hydrocarbons. THC = total hydrocarbons.

**Destruction Efficiency = $\frac{V_{in} - V_{out}}{V_{in}}$ x 100% where, V = THC concentration (in or out), ppmv.

***These solvents are added in small amounts to premixed coatings purchased by Plant C-1. DIBK = diisobutyl ketone, MIBK = methylisobutyl ketone.

†Extended process down time not included in this period.

††Three of eight inlet data points for this period exceeded full-scale value of 5145 ppmv.

†††Process problems — unsteady operation.

§Several inlet data points exceeded full-scale value of 12140 ppmv during this period.

TABLE B-3. SUMMARY OF EPA METHOD 25A SAMPLING RESULTS - PLANT C-2

Press Status	Incinerator Status	Average Inlet THC Concentration, ppmc	Average Outlet THC Concentration, ppmc	Average Inlet THC Mass Rate(M_i), kgC/hr[a] (lbC/hr)	Average Outlet THC Mass Rate(M_o), kgC/hr[a] (lbC/hr)	THC Destruction Efficiency (DE), %[b]
On	On	1020	169	1.70 (3.74)	0.319 (0.702)	81.2
Off	On	187	135	0.312 (0.687)	0.255 (0.561)	18.3
Weighted Average[c]		714	157	1.19 (2.62)	0.295 (0.650)	75.2

[a] $M = \left(\frac{(C)\ \text{mole C}}{\text{mole gas}}\right)\left(\frac{12\ \text{lb C}}{\text{mole C}}\right)\left(\frac{\text{mole gas}}{378\ \text{ft}^3}\right)\left(\frac{(Q)\ \text{ft}^3}{\text{min}}\right)\left(\frac{60\ \text{min}}{\text{hr}}\right)\left(\frac{0.454\ \text{kg}}{\text{lb}}\right)$, where C = THC concentration, ppmc
Q = gas flow rate, scfm (wet).

[b] $DE = \frac{M_i - M_o}{M_i} \times 100\%$.

[c] Press on 63.3% of time; press off 36.7% of time.

TABLE B-4. SUMMARIZED EPA METHOD 25A RESULTS FOR INCINERATOR 20 - PLANT C-3

Test Number	Catalyst Temperature Setpoint, °C (°F)	Operating Conditions/Hours[a]	Product ID	Web Speed, m/sec (fpm)	Inlet THC Concentration (C_i), ppmc	Outlet THC Concentration (C_o), ppmc	Catalyst ΔT, °C (°F)	Destruction[c] Efficiency, %
1	371 (700)	Steady-state/5.5	A	6.65 (1310)	1240	241	7 (12)	80.1
		Unsteady-state/4.1		-b	280	98	-	65.0
		Total/9.6		-	830	180	-	78.3
2	385 (725)	Steady-state/6.7	A	6.60 (1300)	1530	248	9 (16)	83.8
		Unsteady-state/1.7		-	168	30	-	82.1
		Total/8.4		-	1250	204	-	83.7
3	399 (750)	Steady-state/3.0	A	7.11 (1400)	1690	227	10 (18)	86.6
		Unsteady-state/1.6		-	177	95	-	46.3
		Total/4.6		-	1160	181	-	84.4

[a] Does not include time periods in which (1) instruments were undergoing calibration, (2) the incinerator was shut down, or (3) instrument malfunctions invalidated the data.

[b] Value is highly variable and, therefore, not calculated or monitored.

[c] Destruction Efficiency = $\frac{M_{in} - M_{out}}{M_{in}} \times 100\%$, where

$$M = \text{THC mass rate} = \frac{(V)\ \text{moles C}}{10^6\ \text{moles gas}} \times \frac{12\ \text{gC}}{\text{mole C}} \times \frac{\text{mole gas}}{0.0237\ \text{m}^3\text{gas}} \times \frac{(Q)\ \text{m}^3\text{gas}}{\text{sec}}$$

V = THC concentration, ppmc (in or out)

Q = gas flow rate, Nm^3/sec (in or out)

TABLE B-5. SUMMARIZED EPA METHOD 25A RESULTS FOR INCINERATOR 21 - PLANT C-3

Test Number	Catalyst Temperature Setpoint, °C (°F)	Operating Conditions/Hours[a]	Product ID	Web Speed, m/sec (fpm)	Inlet THC Concentration (C_i), ppmc	Outlet THC Concentration (C_o), ppmc	Catalyst ΔT, °C (°F)	Destruction Efficiency, %[c]
1	371 (700)	Steady-state/5.5	B	5.74 (1130)	4030	165	63 (114)	95.9
		Unsteady-state/3.0		-[b]	807	55	-	93.2
		Total/8.5		-	2900	126	-	95.7
		Steady-state/1.7	C	5.59 (1100)	1370	90	22 (40)	93.4
		Unsteady-state/0.9		-	296	32	-	89.2
		Total/2.6		-	983	70	-	92.9
2	385 (725)	Steady-state/4.9	C	5.33 (1050)	1670	92	29 (53)	94.5
		Unsteady-state/0.9		-	136	27	-	80.0
		Total/5.8		-	1440	82	-	94.3
		Steady-state/5.4	D	6.10 (1200)	1910	100	28 (50)	94.7
		Unsteady-state/0.7		-	604	74	-	87.7
		Total/6.1		-	1760	94	-	94.6
3	399 (750)	Steady-state/4.4	E	5.94 (1170)	2910	115	42 (75)	96.0
		Unsteady-state/1.0		-	729	29	-	96.0
		Total/5.4		-	2500	99	-	96.0
		Steady-state/1.9	F	4.17 (820)	2070	84	39 (71)	95.9
		Unsteady-state/7.3		-	933	44	-	95.3
		Total/9.2		-	1170	52	-	95.6

[a]Does not include time periods in which (1) instruments were undergoing calibration and (2) the incinerator was shut down.

[b]Value is highly variable and, therefore, not calculated or monitored.

[c]Destruction Efficiency = $\frac{M_{in} - M_{out}}{M_{in}} \times 100\%$, where

$$M = \text{THC mass rate} = \frac{(V)\ \text{moles C}}{10^6\ \text{moles gas}} \times \frac{12\ \text{gC}}{\text{mole C}} \times \frac{\text{mole gas}}{0.0237\ \text{m}^3\text{gas}} \times \frac{(Q)\ \text{m}^3\ \text{gas}}{\text{sec}}$$

V = THC concentration, ppmc (in or out)

Q = gas flow rate, Nm^3/sec (in or out)

TABLE B-6. SUMMARY OF METHOD 25A SAMPLING RESULTS - PLANT C-4

Date	Monitoring Period	Length of Monitoring Period (min)	Product ID	Average Gas Temperature °C (°F) Catalyst Inlet	Average Gas Temperature °C (°F) Catalyst Outlet	Inlet THC Concentration (ppmv-C)[a]	Outlet THC Concentration (ppmv-C)[a]	Average Destruction Efficiency[d]
1/31/83	2139 - 2252	73	A	318 (605)	382 (720)	7,562	408	94.0
2/1/83	0727 - 0809	45	A	318 (605)	366 (690)	6,822[b]	385	93.8[b]
2/1/83	1929 - 2101	94	A	335 (635)	363 (685)	2,545[c]	251	88.8[c]
2/3/83	1300 - 1911	323	B	335 (635)	396 (745)	5,998	601	88.9
2/3/83	2051 - 2301	130	B	335 (635)	385 (725)	5,484	564	88.7
2/4/83	1017 - 1117	61	B	332 (630)	385 (725)	4,725	546	87.3
2/4/83	1210 - 1222	12	B	335 (635)	354 (670)	4,198	530	86.1
2/4/83	1440 - 1710	152	B	368 (695)	432 (810)	6,060	495	91.0
2/4/83	1824 - 2043	142	B	363 (685)	443 (830)	6,546[b]	531	91.0[b]
2/4/83	2044 - 2213	90	B	377 (710)	454 (850)	6,403[b]	445	92.3[b]
2/4/83	2217 - 2242	25	B	393 (740)	460 (860)	6,435[b]	440	92.4[b]
2/4/83 - 2/5/83	2259 - 0101	123	B	391 (735)	452 (845)	5,275	363	92.4
2/5/83	0102 - 0158	57	B	371 (700)	432 (810)	5,526[b]	500	90.0[b]
2/5/83	0159 - 0310	71	B	335 (635)	402 (755)	5,618	592	88.4
2/5/83	0319 - 0534	138	B	335 (635)	407 (765)	6,418	687	88.2
2/5/83	0608 - 1113	264	B	332 (630)	388 (730)	4,714	501	88.3

[a] ppmv-C = parts per million by volume as carbon quanitated against propane.

[b] Unsteady catalyst inlet temperature.

[c] Unsteady process operation.

[d] Destruction Efficiency = $\frac{Min - Mout}{Min}$ x 100%, where

$$M = \text{THC Mass Rate} = \frac{(V)\ \text{moles C}}{10^6\ \text{moles gas}} \times \frac{12\text{gC}}{\text{mole C}} \times \frac{\text{mole gas}}{0.024\text{m}^3\ \text{gas}} \times \frac{(Q)\ \text{m}^3\ \text{gas}}{\text{sec}}$$

V = THC concentration (inlet or outlet), ppmv

Q = gas glowrate (inlet or outlet), Nm^3/sec

TABLE B-7. SUMMARY OF METHOD 25A SAMPLING RESULTS - PLANT C-5

Date	Monitoring Period	Minutes of Data Collected	Inlet THC Concentration ppmv-C	Outlet THC Concentration ppmv-C	Destruction Efficiency (%)	Catalyst Inlet Temperature °C (°F)	Catalyst Outlet Temperature °C (°F)
2/15/83	18:17-20:01	105	8954	1876	77.8[a]	396 (745)	485 (905)
2/15-2/16/83	20:07-03:57	470	8908	2194	73.9	407 (765)	499 (930)
2/16/83	03:58-06:04	127	NA	2452	NA	407 (765)	499 (930)
	06:06-08:03	118	NA	2651	NA	402 (755)	488 (910)
	08:04-10:13	118	8710	2422	70.5	410 (770)	502 (935)
	16:48-19:23	162	8535	1404	82.5[b]	379 (715)	499 (930)
2/16-2/17/83	19:31-09:17	826	8720	1591	80.6	393 (740)	507 (945)
2/17/83	09:24-11:23	120	8299	1666	78.7	399 (750)	513 (955)
	11:28-12:33	63	8616	1716	78.9[a]	457 (855)	554(1030)
2/17-2/18/83	18:08-08:43	841	9021	1529	82.0	493 (920)	577(1070)
2/18/83	08:43-10:17	93	8825	1459	82.5[a]	488 (910)	568(1055)
	12:44-14:46	123	8862	1374	83.5	468 (875)	554(1030)
	17:04-21:15	247	9175	1564	81.1[a]	468 (875)	549(1020)
	21:16-22:12	57	8637	1541	81.1	418 (785)	502 (935)
2/18-2/19/83	22:13-00:09	117	8645	1669	79.5	402 (755)	482 (900)
2/19/83	00:10-00:40	31	8841	1916[c]	77.0	399 (750)	479 (895)
	00:41-02:15	95	4638	1526	65.1[d]	413 (775)	457 (855)
	02:16-05:18	183	5047	1155	75.7[d]	418 (785)	454 (850)
	05:19-08:49	204	8480	1605	79.9[e]	399 (750)	477 (890)

NA - Data not available.

[a]Temperature unsteady during a portion of the monitoring period - typically this was 50-100% of monitoring time.

[b]Data collected during period immediately following catalyst changeout. Catalyst not at steady state.

[c]Outlet concentration increased during this period from 1700-2000 ppm. Suspected cause of increase is process upset just prior to shutdown of ½ of coater.

[d]Only half of wire coater operating during this period.

[e]Process not back to steady state after half of coater was down.

TABLE B-8. SUMMARY OF EPA METHOD 25A SAMPLING RESULTS - PLANT C-6

Date	Monitoring Period	Length of Monitoring Period	Product ID	Main Inlet THC Concentration[a] (ppmv)	Floor Sweep THC Concentration[a] (ppmv)	Outlet THC Concentration[a] (ppmv)	THC Mass Flowrates Inlet kg/hr (lb/hr)	THC Mass Flowrates Outlet kg/hr (lb/hr)	Destruction Efficiency[k] (%)	Catalyst Inlet Temperature °C (°F)	Catalyst Outlet Temperature °C (°F)
2/25/83	17:02-18:06[b]	65	F	8006	--	201	71.2 (157)	2.0 (4.3)	97.3	285 (545)	393 (740)
2/25/83	20:53-21:34[b]	42	F	8811	--	199	78.0 (172)	1.9 (4.2)	97.6	285 (545)	396 (745)
2/28/83	07:25-08:11[b]	47	B	11125	--	263	98.0 (216)	2.5 (5.6)	97.4	285 (545)	416 (780)
2/28/83	14:06-15:06	66	B	10947	--	291	96.2 (212)	2.8 (6.2)	97.1	285 (545)	427 (800)
2/28/83	15:39-17:37	118	B	10830	2382[c]	294	94.8 (209)	2.8 (6.2)	97.0	282 (540)	416 (780)
2/28/83	18:13-19:40	87	B	10926	2596[d]	305	96.2 (212)	3.0 (6.5)	97.0	277 (530)	418 (785)
2/28/83	19:47-20:57	72	B	10970	--	290	96.6 (213)	2.8 (6.2)	97.1	277 (530)	418 (785)
2/28/83	21:47-22:04	18	B	10217	--	291	90.3 (199)	2.8 (6.2)	96.9	277 (530)	418 (785)
3/1/83	13:58-15:00	63	A	11514	--	287	101 (223)	2.8 (6.1)	97.3	274 (525)	421 (790)
3/1/83	17:11-18:24	74	A	12864	--	295	112 (248)	2.9 (6.3)	97.5	285 (545)	443 (830)
3/1/83	19:56-20:59	64	A	12504[e]	2592[f]	286	109 (241)	2.8 (6.1)	97.5	282 (540)	435 (815)
3/1/83	22:06-23:36	91	A	12303	--	289	108 (238)	2.8 (6.2)	97.4	285 (545)	438 (820)
3/2/83	00:31-02:30	93	A	12505	--	293	109 (241)	2.8 (6.2)	97.4	285 (545)	446 (835)
3/2/83	04:38-06:38	123	B	10183	--	280	89.8 (198)	2.7 (5.9)	97.0	282 (540)	421 (790)
3/2/83	07:13-08:50	100	B	11200	--	293	98.4 (217)	2.8 (6.2)	97.1	282 (540)	438 (820)
3/2/83	09:33-11:14	102	B	12358	--	250	108 (239)	2.4 (5.3)	97.8	338 (640)	474 (885)
3/2/83	12:32-14:02	91	B	10122	--	218	89.4 (197)	2.1 (4.7)	97.6	341 (645)	457 (855)
3/2/83	19:37-21:21	64[g]	C	8806	--	286	78.0 (172)	2.8 (6.1)	96.5	274 (525)	424 (795)
3/3/83	09:32-11:04	93	D	6485	--	269	58.1 (128)	2.6 (5.7)	95.5	252 (485)	335 (635)
3/3/83	11:44-12:15	32	D	6222	--	272	55.8 (123)	2.6 (5.8)	95.3	254 (490)	341 (645)
3/3/83	15:56-17:11	76	D	6542	2452[h]	248	58.5 (129)	2.4 (5.3)	95.9	288 (550)	374 (705)
3/3/83	20:15-00:03	217	B	8794	2888[i]	245	78.0 (172)	2.4 (5.3)	97.0	341 (645)	443 (830)
3/4/83	00:51-01:57	67	B	9594	--	298	84.8 (187)	2.9 (6.3)	96.6	279 (535)	407 (765)
3/4/83	02:43-05:06	144	B	9467	--	299	83.9 (185)	2.9 (6.4)	96.5	285 (545)	407 (765)
3/4/83	05:40-07:55	136	B	9212	--	295	81.7 (180)	2.9 (6.3)	96.5	285 (545)	407 (765)
3/4/83	17:32-20:21	95	E	11244	2372[j]	279	98.4 (217)	2.7 (5.9)	97.3	277 (530)	435 (815)
3/4/83	21:27-22:09	43	E	11101	--	250	97.1 (214)	2.4 (5.3)	97.5	307 (585)	424 (795)

[a] ppmv - parts per million by volume as carbon C_1 quanitated against propane. Gas sampling locations are shown in Figure 5-1.
[b] Data collected during these periods represent a SLA residence time of approximately 0.036 sec. All other data represent a residence time of 0.029 sec.
[c] Based on 6 minutes of data analysis.
[d] Based on 63 minutes of data analysis.
[e] Inlet concentration based on 15 minutes monitoring during this period.
[f] Based on 42 minutes of monitoring.
[g] Process down period not included in average.
[h] Based on 9 minutes.
[i] Based on 63 minutes of data.
[j] Based on 39 minutes of data.
[k] Destruction Efficiency $= \frac{M_{in} - M_{out}}{M_{in}} \times 100\%$, where

$$M - \text{THC Mass Rate} = \frac{(V) \text{ moles C}}{10^6 \text{ moles gas}} \times \frac{12 \text{gC}}{\text{mole C}} \times \frac{\text{mole gas}}{0.024 \text{ m}^3 \text{ gas}} \times \frac{(Q) \text{ m}^3 \text{ gas}}{\text{sec}}$$

V - THC concentration (inlet or outlet); ppmv

Q - gas flowrate (inlet or outlet), Nm^3/sec

Appendix C

Equivalent Inlet THC Concentration for Plant C-6

The method used to calculate an "equivalent" inlet THC concentration for Plant C-6 is presented in this section. An estimate of the concentration was required for the comparative analysis presented in Section 6.3. The calculation of inlet THC mass rate for the Plant C-6 incinerator will be discussed first. The equivalent concentration calculation method will follow.

The total inlet THC mass rate of Plant C-6 was determined as the sum of THC mass rates in three gas streams: main inlet, floor sweep, and combustion air. These gas streams and the corresponding sampling locations are shown in Figure C-1. The total inlet THC mass rate was calculated as follows:

$$M_{tot} = \Sigma M_i$$

$$M = \frac{(V)\ \text{moles}}{10^6\ \text{moles gas}} \times \frac{12\text{gC}}{\text{mole C}} \times \frac{\text{moles gas}}{0.024\ \text{m}^3\text{gas}} \times \frac{(Q)\text{m}^3\ \text{gas}}{\text{sec}}$$

where M = THC mass rate (main inlet, floor sweep or combustion air), g_C/sec

Q = gas flowrate (main inlet, floor sweep or combustion air), Nm^3/sec

V = THC concentration (main inlet, floorsweep or combustion air), ppmv

Although the actual inlet THC concentration at the incinerator inlet was not measured, an equivalent concentration can be estimated from the total inlet mass rate:

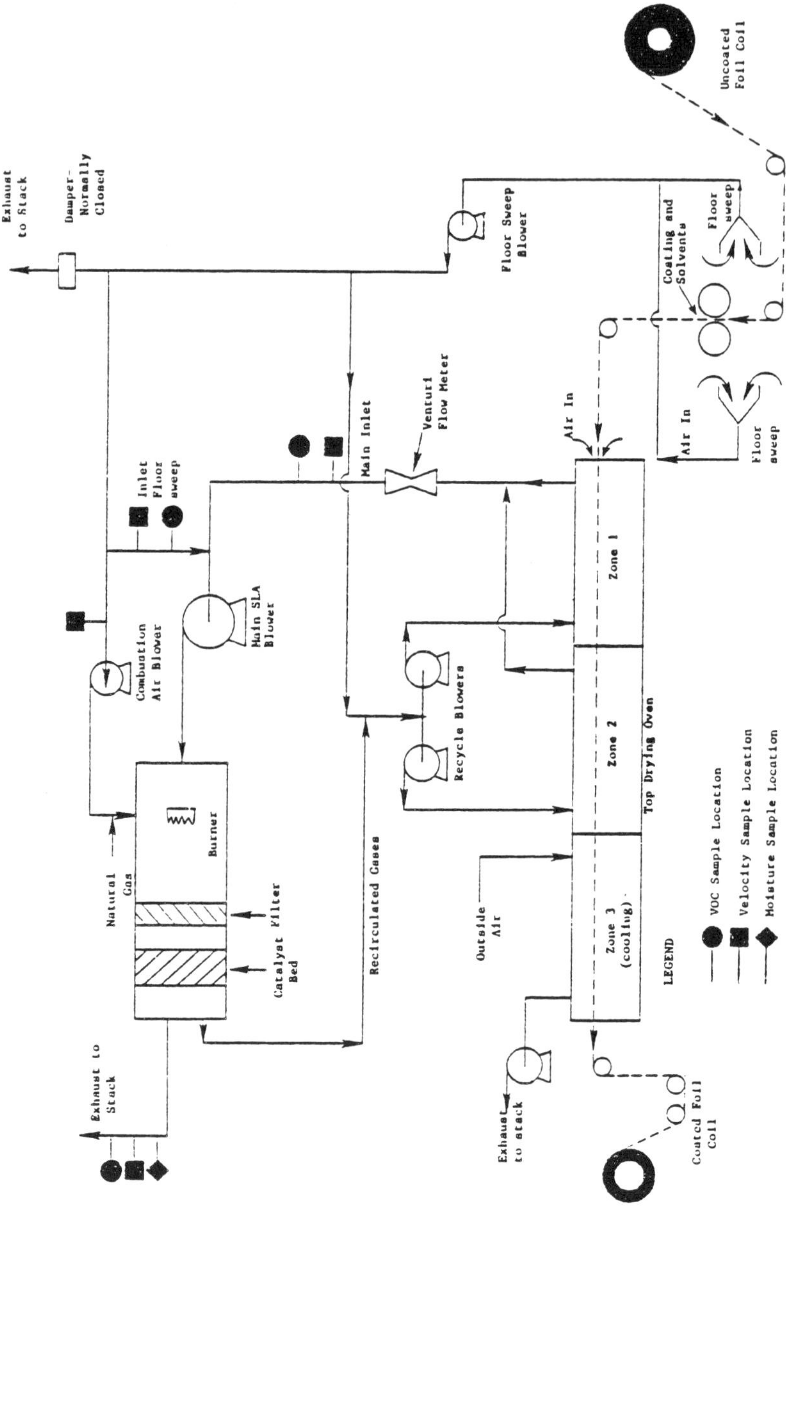

Figure C-1. Process flow diagram showing sample points for Plant C-6.

$$V_{in}, EQ = M_{tot} \times 10^6 \text{ moles gas} \times \frac{\text{mole C}}{12 \text{ gC}} \times \frac{0.024 \text{ m}^3 \text{ gas}}{\text{moles gas}}$$

$$\times \frac{1}{(Q_{tot}) \text{ m}^3\text{gas/sec}}$$

where V_{in}, EQ = equivalent inlet THC concentration, ppmv

Q_{tot} = sum of main inlet, floorsweep and combustion air gas flowrates, Nm^3/sec.

The inlet concentrations presented for Plant C-6 in Section 6.3 were calculated using the above equations. The total inlet THC mass flowrates used for the calculations were taken directly from the Plant C-6 test report.